Quantitative Genetics and Its Connections with Big Data and Sequenced Genomes

Quantitative Genetics and Its Connections with Big Data and Sequenced Genomes

Charles J Mode

Drexel University, USA

NEW JERSEY • LONDON • SINGAPORE • BEIJING • SHANGHAI • HONG KONG • TAIPEI • CHENNAI • TOKYO

Published by

World Scientific Publishing Co. Pte. Ltd.
5 Toh Tuck Link, Singapore 596224
USA office: 27 Warren Street, Suite 401-402, Hackensack, NJ 07601
UK office: 57 Shelton Street, Covent Garden, London WC2H 9HE

Library of Congress Cataloging-in-Publication Data
Names: Mode, Charles J., 1927–
Title: Quantitative genetics and its connections with big data and sequenced genomes / Charles J. Mode, Drexel University, USA.
Description: New Jersey : World Scientific, 2016. | Includes bibliographical references and index.
Identifiers: LCCN 2016028746| ISBN 9789813140677 (hardcover : alk. paper) | ISBN 9813140674 (hardcover : alk. paper) | ISBN 9789813140684 (pbk. : alk. paper) | ISBN 9813140682 (pbk. : alk. paper)
Subjects: LCSH: Quantitative genetics. | Nucleotide sequence. | Molecular biology.
Classification: LCC QH452.7 .M63 2016 | DDC 572.8/633--dc23
LC record available at https://lccn.loc.gov/2016028746

British Library Cataloguing-in-Publication Data
A catalogue record for this book is available from the British Library.

Printed in Singapore

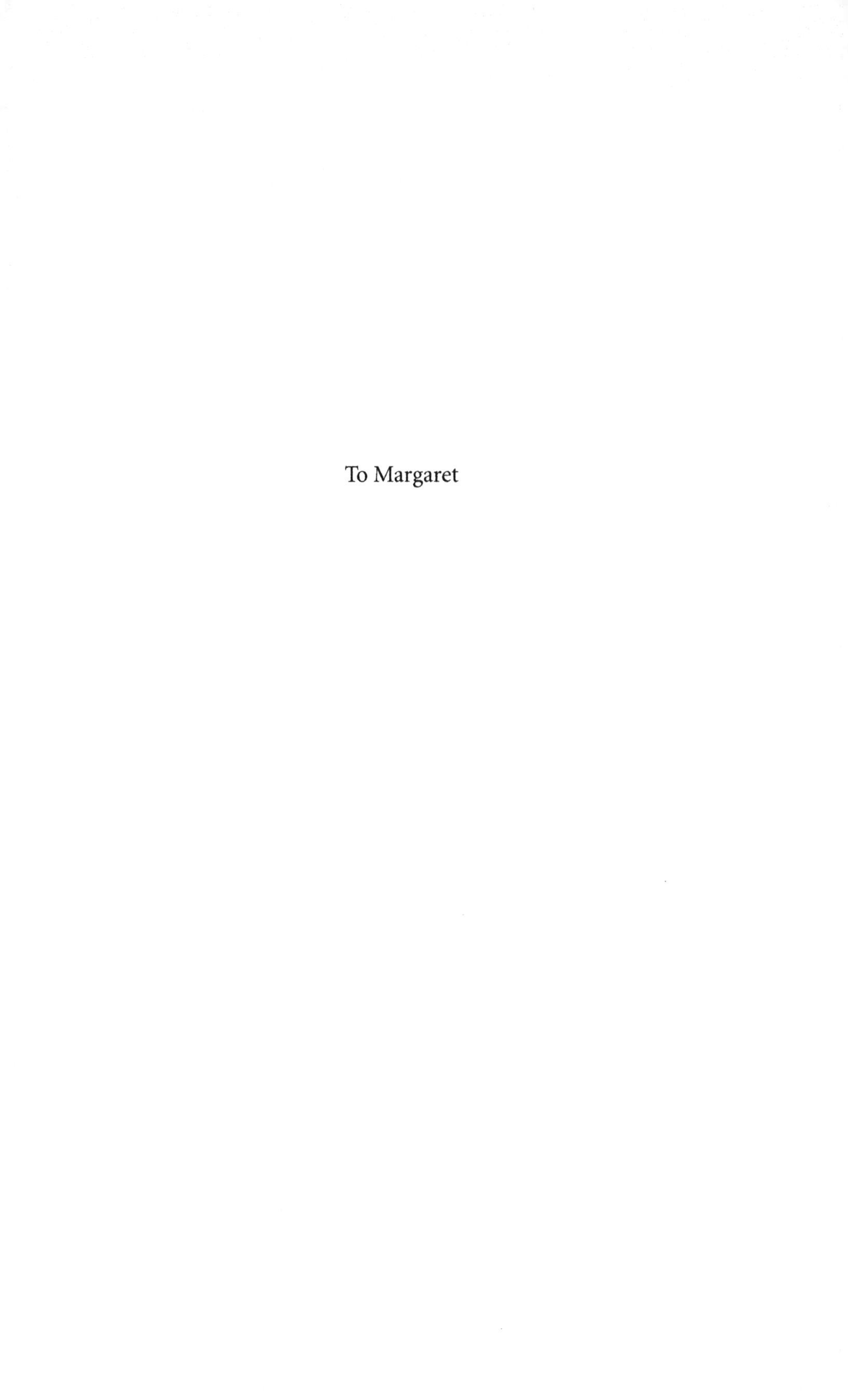

To Margaret

Preface

The period 1957 to 1966 was a time in which the author worked with variance component models that were widely used in quantitative genetics, but after 1966 his attention was turned to other subjects, such as evolution, branching processes, and the teaching of probability and statistics. Recently, however, the author became aware that the genomes of large samples of patients with Alzheimer's disease had been sequenced. Furthermore, there was statistical evidence those 11 regions of the human genome were involved in the quantitative expression traits associated with patients with Alzheimer's disease. Subsequently, it was recognized that it would be possible obtain working definitions of 11 loci corresponding to the 11 regions in the human genome, with at least two alleles at each locus defined in terms of markers such as the presence and absence of nucleotide substitutions.

This recognition, in turn, led to awareness that not only variance components but also the effects that were a basis for their definitions could be estimated directly rather than by the indirect methods, based on analysis of variance and covariance procedures, that have been widely used in quantitative genetics. Briefly, this small book, which contains five chapters, is a short and incomplete history of quantitative genetics based on the experience of the author over a period of at least 40 years.

Presented in Chapter 1 is a paper that was developed and written during the period 1956 to 1958, in which attention was focused on analysis of variance and covariance procedures that were used extensively during that period in connection with the estimation of the genetic and variance and covariance components making up models of quantitative genetics. After a brief discussion of the circumstances that gave rise to the paper, prospective and retrospective views of the paper are given, using properties of matrices. Chapter 2 presents a paper on testing a hypothesis as to whether a major gene was sufficient to explain a quantitative trait that arose in connection with work on human blood. Again, after a brief discussion of the circumstances that led to the paper, a revised version of the paper on fitting models to data, which was published in the 1970s, is republished in a revised form.

The content of Chapter 3 is a reproduction of a recently published theoretical paper containing a detailed account of the direct estimation of effects and variance components for the case of one quantitative trait. In this paper, effects are defined mathematically. The case of one autosomal locus is considered, as well as cases of multiple loci. Similarly, Chapter 4 is a reproduction of a recently published paper, on pleiotropism and the direct estimation of variance and covariance components and effects that arise when several quantitative traits are under consideration. To conclude the book, Chapter 5 contains a republished account, based on artificial data, of the estimation as well as tests of statistical significance as to whether the estimated effects are indeed not equal to zero for the case of one autosomal locus with two alleles. Monte Carlo simulation procedures were employed to conduct these tests as well as to estimate two types of *p* values, which were used to judge the statistical significance of estimated effects. The last three chapters contain ideas designed for investigating the genetics of some quantitative in big data sets, i.e., large samples and sequenced genomes.

Acknowledgments

Many thanks are due to Dr. Candace Sleeman for her diligent work in constructing the table of contents and the subject index. An innovative aspect of this work was the use of software that picked up the chapter titles as well as the titles of the sections in each chapter with their page numbers. Software was also used to select words and phrases in the subject index along with the page number for each word or phrase.

Contents

CHAPTER 1

An Example of an Application of Variance and Covariance Models in Quantitative Genetics in the 1950s

BACKGROUND

During the period 1956 to 1957, I had a postdoctoral position in the Department of Experimental Statistics, North Carolina State University at Raleigh. My supervising professor was Dr. H. F. Robinson, who, along with his associates, had done extensive research on the quantitative genetics of corn (Zea maize). Usually, measurements were taken on two or more traits, and the data were summarized in terms of variance–covariance tables. The 1950s were a time before high level programming languages, such as FORTRAN, had been developed. Consequently, the programs used to compute the variance–covariance tables must have been written in a very primitive programming language, which very likely involved extensive and tedious work in writing and editing the computer code. Dr. Robinson was part of a team that was working on quantitative genetics during the 1950s. The work of this team was recognized by an international community of geneticists. It was the international reputation of this team that had led me to apply for a postdoctoral position after I had completed my Ph.D. in genetics in 1956, with study and work at the University of California, Davis campus as well as the Berkeley campus. At that time, Dr. Robinson was a member of the Department of Experimental Statistics at North Carolina State University.

Another influential member of the team in this department was Dr. R. E. Comstock, who was very knowledgeable about the experimental designs underlying the variance and covariance models used to summarize the data that were generated in field experiments with inbred lines of corn. Dr. Robinson had suggested that I work with the summarized data in one of the variance–covariance tables in an effort to gain a more in-depth understanding of the mathematics underlying the use of such tables when two quantitative traits were under consideration. The two traits that were studied in the experiment were ear and plant height. After I had read the paper by Comstock and Robinson (1948) as a step toward developing some ideas as to how to get my research underway, I asked Dr. Comstock why the standard errors of the parameters estimated in the 1948 paper were not listed. In response to this question he told me that the theory underlying the computation of such standard errors had not been developed. Consequently, I decided that the development of a computational procedure for estimating these standard errors would be one of the goals of my investigation into the mathematics underlying the variance–covariance tables.

My next step in developing ideas as to how one can estimate the standard errors of nonlinear functions of estimated parameters was to consult David Hurst, who was a graduate student pursuing a Ph.D. in the Department of Experimental Statistics at North Carolina State University. Hurst had a very good background in mathematics and suggested that I expand the nonlinear functions into series and retain only the first order terms so that an approximate standard error could be estimated for all of the nonlinear functions of estimated parameters. But another event occurred shortly after the conversations with Hurst that delayed further work on my research project.

In either late December of 1956 or early January of 1957, the head of the Agronomy Department at the University of Florida contacted the Department of Experimental Statistics at North Carolina State University, informing them that an interim position was available for teaching a course in experimental design to graduate students in agronomy in the spring semester of 1957. A few years earlier, in 1952 and 1953, I had taken a course in experimental design at Kansas State University in Manhattan from a statistician in the Mathematics Department, while pursuing a Master's degree in agronomy and genetics. I found that the content of the course was compelling and useful, so I worked hard on learning how methods of experimental design could be used in my future research. Moreover, while pursuing a Ph.D. degree in genetics at the University of California at Davis, I had acquired sufficient knowledge of the mathematics underlying experimental design to feel confident that I could teach the course. Consequently, I told the statisticians in the Department of Experimental Statistics that I was able to teach the course, and within a few days I was on my way to the University of Florida with the aim of teaching a course in experimental design to graduate students in agronomy.

Teaching the course in experimental design was a very rewarding experience for me. Many of the students in the class were from foreign countries, including Cuba and India. Several of the students from Cuba were older than me, and a student from India had a very inquisitive mind. Most of the content of the course involved analysis of variance procedures, and when I derived some mathematical formulas that were part of the subject, the older students from Cuba sent me signals that they understood the mathematics. But the inquisitive student from India was surprised to see that these formulas could be derived. Prior to taking the course, he thought that R. A. Fisher, a famous English statistician and geneticist, had set them down and they were accepted as laws among biologists and others who used them. It was very rewarding to me that I, as a teacher, was able to give the student insight into the mathematical nature of statistics, and that it was not only the thoughts of one famous man but also a community of scholars that had an understanding of its nature and were contributing to the substance of the field of statistics. After deriving much satisfaction from teaching the course in experimental design, I came to the realization that I wanted teaching to be part of my career.

Sometime in early July, I had returned to the Department of Experimental Statistics at North Carolina Sate University and resumed work on my research project as well as helping with field experiments designed to study the quantitative genetics of corn. In my spare time, I continued my conversations with David Hurst on the mathematics underlying my research and we also talked about more general aspects of mathematics. Sometime in August of 1957, I received an unexpected letter from Dr. Hurst, who was the father of David Hurst and head of the Department of Mathematics at Montana State University. Much to my surprise, I was offered a tenure track position in the Department of Mathematics, with the understanding that I would be a member of the department but also a consultant to the College of Agriculture. By the time of the offer, I was confident that I that could function in a mathematics department so I accepted Dr. Hurst's offer, which was to change the direction of my career.

It was as a member of the Department of Mathematics at Montana State University that the research into the problem on quantitative genetics was completed along with the writing of the paper which was accepted for publication in *Biometrics*; see Mode and Robinson (1959). The knowledge of multivariate statistics acquired in the department greatly expedited the completion of research that culminated in the writing of the first draft of the paper. Among the novel features of this paper were formulas for computing the standard errors of estimates for all parameters in the paper. Dr. Robinson was very pleased with this paper and republished it in a volume of well-known papers on quantitative genetics, which was published by the Department of Genetics, North Carolina State University. The two traits studied in this paper were ear and plant height in corn.

As the content of the paper was used by other investigators in quantitative genetics, it became apparent that there were flaws in the estimation procedure described in this paper. Consequently, rather than republishing the content of the paper, a decision was made to have a retrospective look at the methods used in the paper as well as a perspective view on how the ideas of the paper may also be used in the present age of sequenced genomes.

A RETROSPECTIVE AND PROSPECTIVE VIEW OF THE PAPER BY MODE AND ROBINSON (1959)

Many years have passed since the mid-1950s, when I wrote the paper on pleiotropism and the genetic variance and covariance, which was published in *Biometrics* in 1959 with Dr. Robinson as the coauthor. In the 1950s, when I worked on this paper, I was primarily a geneticist interested in plant breeding. During the time I was taking courses while working on my Ph.D. in genetics, however, I was advised by Dr. Robert Allard, a plant breeder in the Department of Agronomy at the University of California, Davis, that if I was serious about increasing my ability to make mathematics an integral part of my prospective career in teaching and research, then I would need to take courses in the more difficult mathematics that were part of the curricula for students in physics, chemistry and engineering, rather than the simpler mathematics courses designed for biologists. I followed Dr. Allard's advice and by the time I was awarded a Ph.D. in genetics at the University of California, Davis, in the spring of 1956, I had completed a semester course in analytic geometry, several semester courses in calculus and a two semester course in mathematical statistics, offered by the Department of Mathematics, University of California, Davis. In all these courses, I had achieved A's, so that when I arrived at North Carolina State University in the early summer of 1956 to begin work in my position, I felt confident that I would be able to work in statistical genetics under the direction of Dr. Robinson.

As mentioned before, early in the fall of 1957 I accepted a tenure track position in the Department of Mathematics at Montana State University, and as a member of this department I was advised to read papers and books on mathematical statistics. Early in 1958 a book titled *An Introduction to Multivariate Statistical Analysis*, by T. W. Anderson, was published; Anderson (1958). Even though much of the content of the book was beyond my level of expertise, I realized that some results on an extension of analysis of variance procedures to the multivariate case would be useful as I continued work on my research paper with Dr. Robinson. With a view to understanding the goals that were part of the motivating forces underlying the writing of the paper, the introduction and theory as stated in the 1959 paper are partially reproduced below, along with a few corrections and remarks that help clarify the content of the paper.

1. Introduction and Theory

In classical genetics many genes are known to have manifold effects, i.e. the gene seems to affect unrelated characters. An example of such a gene is the "vestigial gene" in *Drosophila*, which affects not only the bristles and wings but also fecundity. Examples of similar genes in other organisms abound in the literature. When a gene has manifold effects, its action is called "pleiotropic".

In quantitative genetics, just as in qualitative genetics, it is easy to conceive of a single gene affecting many characteristics. We will, therefore, consider a random mating population with respect to one segregating locus at which there are an arbitrary number of alleles, and suppose that two characters, X and Y, are observed. If the two genes act pleiotropically, to each genotype, A_iA_j, there correspond two genotypic values, X_{ij} for character X and Y_{ij} for character Y.

Let the genotypic values X_{ij} and Y_{ij} be deviations from their respective means. Then, if p_i and p_j are their respective frequencies,

$$\sum_{i,j} p_i p_j X_{ij} = \sum_{ij} p_i p_j Y_{ij} = 0, \tag{1.1}$$

where the summation extends over all alleles. We now introduce the following definitions. The additive effects of the genes with respect to the two characters in question are defined as

$$\begin{aligned} \alpha_i^* &= X_{io} = \sum_j p_i X_{ij}, \\ \alpha_j^* &= X_{oj} = \sum_i p_i X_{ij} \end{aligned} \tag{1.2}$$

for character X and

$$\begin{aligned} \alpha_i &= Y_{oi} = \sum_i p_i Y_{ij}, \\ \alpha_j &= Y_{oi} = \sum_i p_i Y_{ij} \end{aligned} \tag{1.3}$$

for character Y.

Because of the symmetry of the genetic mechanism, $\alpha_i^* = \alpha_j^*$ and $\alpha_i = \alpha_j$ if $i = j$. The dominance deviations are defined as

$$\delta_{ij}^* = X_{ij} - \alpha_i^* - \alpha_j^* \tag{1.4}$$

for character X and

$$\delta_{ij} = Y_{ij} - \alpha_i - \alpha_j \tag{1.5}$$

for character Y.

The components of variance and covariance are defined as follows. The total genetic variances and covariance are defined as

$$\sigma_{G^*}^2 = \sum_{(i,j)} p_i p_j X_{ij}^2,$$

$$\sigma_G^2 = \sum_{(i,j)} p_i p_j Y_{ij}^2$$

and

$$\sigma_{GG^*} = \sum_{(i,j)} p_i p_j X_{ij} Y_{ij}. \tag{1.6}$$

The additive genetic variances and covariance are defined as

$$\sigma^2_{A^*} = \sum_i p_i \alpha_i^{*2} + \sum_j p_j \alpha^{*2},$$
$$\sigma^2_A = \sum_i p_i \alpha_i^2 + \sum_j p_j \alpha^2$$

and

$$\sigma_{AA^*} = \sum_i p_i \alpha_i \alpha_i^*. \tag{1.7}$$

Rather than naming the components of variance and covariance, in Chapters 3 and 4 the expressions in (1.7) are referred to as the additive variances and covariance. If the population is in a Hardy–Weinberg equilibrium, then it can be shown that

$$\sigma^2_{G*} = \sigma^2_{A*} + \sigma^2_{D*},$$
$$\sigma^2_G = \sigma^2_A + \sigma^2_D,$$
$$\sigma_{GG^*} = \sigma_{AA*} + \sigma_{DD^*}. \tag{1.8}$$

It is thus possible to speak of the genetic covariance just as we speak of the additive and dominate of the genetic variance for the case of one trait.

The above arguments may be generalized in a natural way to random mating populations in which $n > 1$ loci are segregating with an arbitrary number of alleles at each locus. The reader is referred to Kempthorne (1957) and elsewhere for a partition of the genetic variance in such a population into the additive, dominant and epistatic components. But, in this paper, attention will be confined to one locus.

From the above components of the genetic variances and covariances, a number of parameters may be defined which, when estimated, throw light on the nature of the underlying genetic mechanisms. The underlying genetic mechanisms causing a linear association between the two characters may be due to pleiotropy, linkage or both. In this paper, the genetic coefficient of correlation is defined by

$$\eta = \frac{\sigma_{AA^*}}{\sigma_A \sigma_{A^*}}. \tag{1.9}$$

Some authors refer to this coefficient as the additive coefficient of correlation. The genotypic correlation coefficient is defined by

$$\varsigma = \frac{\sigma_{GG^*}}{\sigma_G \sigma_{G^*}}. \tag{1.10}$$

Let σ_E^2 denote the environmental variance for character X. Then the phenotypic variance for character X is defined by

$$\sigma_P^2 = \sigma_G^2 + \sigma_E^2. \tag{1.11}$$

The phenotypic variance for character Y and covariance for characters X and Y may be defined similarly. The environmental variance is usually defined in connection with some experimental design, as will be illustrated in a subsequent section of this paper. Given these definitions, the phenotypic correlation is defined as

$$\theta = \frac{\sigma_{PP^*}}{\sigma_P \sigma_{P^*}}. \tag{1.12}$$

A fourth parameter is of interest as a measure of the average degree of dominance, which is defined by

$$\alpha = \sqrt{\frac{2\sigma_D^2}{\sigma_A^2}}. \tag{1.13}$$

For a detailed discussion of the assumptions underlying the estimation of this parameter and its genetic interpretation see Comstock and Robinson (1948).

At the time before the (1959) paper was written, an investigator, in general, would not know the location of the genes and loci involved in the expression of a quantitative trait. But, as indicated in Chapters 3 and 4, some investigators now have access to samples in which the genome of each individual has been sequenced. Consequently, in order to define the effects corresponding to genes under consideration, it will be assumed that the genotype of each individual is known. Consider a genotype denoted by the symbol A_iA_j and let μ_{ij} denote the mean of $n_{ij} \geq 1$ individuals with this genotype.

Let p_i and p_j denote the frequencies of alleles A_i and A_j in the sample. In the recent literature on quantitative genetics, estimated means of genotypes are referred to as genetic values. Let

$$\mu = \sum_{(i,j)} p_i p_j \mu_{ij} \tag{1.14}$$

denote the mean of all genetic values. Then the genetic values defined above expressed in terms of the means μ_{ij} have the form

$$X_{ij} = \mu_{ij} - \mu \tag{1.15}$$

for character X. At the time the (1959) paper was written, the genetic values were defined implicitly as indicated in this equation. But it was not stated that the mean μ was defined in terms of the genotypic distribution as in (1.14).

In Chapters 3 and 4, additive effects were defined as follows. Let

$$\mu_{i\cdot} = \sum_{j} p_j \mu_{ij}. \tag{1.16}$$

Then the additive effect for allele A_i is defined by

$$\alpha_i = \mu_{i\cdot} - \mu \tag{1.17}$$

for all alleles i. The effect that is a measure on intra-allelic interaction is defined by

$$\delta_{ij} = \mu_{ij} - \mu - \alpha_i - \alpha_j \tag{1.18}$$

for all genotypes (i, j). Equivalently,

$$\mu_{ij} = \mu + \alpha_i + \alpha_j + \delta_{ij} \tag{1.19}$$

for all genotypes (i, j). When two or more traits are under consideration, for each trait its genetic value could be expressed in terms of the last equation. In the next section, these ideas will be revisited.

2. Properties of Covariance Matrices

When two or more traits are under consideration, deeper mathematical insights into a formulation may be attained if covariance matrices are a central focus of attention. For example, let σ^2_{G1} and σ^2_{G2} denote the genetic variances of the two traits, and σ_{G12} denote the covariance of the traits. Then the genetic covariance matrix for the two traits has, by definition, the form

$$\Psi_G = \begin{bmatrix} \sigma_{G1}^2 & \sigma_{G12} \\ \sigma_{G21} & \sigma_{G2}^2 \end{bmatrix}, \tag{2.1}$$

where $\sigma_{G12} = \sigma_{G21}$. The transpose of this matrix is, by definition,

$$\begin{bmatrix} \sigma_{G1}^2 & \sigma_{G21} \\ \sigma_{G12} & \sigma_{G2}^2 \end{bmatrix} \tag{2.2}$$

Observe that this matrix has been derived from that in (2.1) by letting the first column of (2.1) be the first row of the matrix in (2.2). Similarly, let the second column of (2.1) be the second row of the matrix in (2.2). The operation resulting in the construction of the matrix in (2.2) is called the transpose of the matrix Ψ_G. Because $\sigma_{G12} = \sigma_{G21}$, it can be seen that the matrix in (2.1) is equal to that in (2.2). Let

$$\Psi_G^T \tag{2.3}$$

denote the transpose of the matrix in Ψ_G. Then, from (2.2), it can be seen that

$$\Psi_G = \Psi_G^T. \tag{2.4}$$

To generalize this case to any $k \times k$ covariance matrix Ψ for $k \geq 2$, let X denote a $k \times 1$ column vector of the random variables $X_1, X_2, \ldots, X_k$ with $E[X_i] = \mu_i$ for $i = 1, 2, \ldots, k$, and let $\boldsymbol{\mu}$ denote a $k \times 1$ column vector of these expectations. Then, in general, the covariance matrix of the k random variables under consideration is defined by

$$E[(X - \boldsymbol{\mu})((X - \boldsymbol{\mu}))^T] = \Psi. \tag{2.5}$$

From this equation, it can be seen that Ψ is symmetric so that

$$\Psi = \Psi^T. \tag{2.6}$$

In working out the properties of covariance matrices, it will be informative to consider the case $k = 2$ and derive a formula for the variance of the random variable

$$Z = a_1X_1 + a_2X_2,$$

where a_1 and a_2 are arbitrary constant numbers. By definition, the variance of Z is

$$\begin{aligned}\mathrm{var}[Z] &= E[(a_1(X_1-\mu_1)+a_2(X_2-\mu_2))^2]\\ &= a_1^2\mathrm{var}[X_1]+a_2^2\mathrm{var}[X_2]+2a_1a_2\mathrm{cov}[X_1,X_2]. \end{aligned} \tag{2.7}$$

For this case, the covariance matrix with respect to the random variables X_1 and X_2 is

$$\boldsymbol{\Psi}_2 = \begin{bmatrix} \mathrm{var}[X_1] & \mathrm{cov}[X_1,X_2] \\ \mathrm{cov}[X_2,X_1] & \mathrm{var}[X_2] \end{bmatrix}, \tag{2.8}$$

where $\mathrm{cov}[X_2,X_1]=\mathrm{cov}[X_1,X_2]$. Let $\boldsymbol{a}^T=(a_1,a_2)$. Then it can be seen that the vector–matrix representation of the var$[Z]$ in (2.7) is

$$\mathrm{var}[Z] = \boldsymbol{a}^T\boldsymbol{\Psi}_2\boldsymbol{a}. \tag{2.9}$$

It is suggested that the reader expand (2.9) in terms of the elements of the vector $\boldsymbol{a}$ and covariance matrix $\boldsymbol{\Psi}_2$ to derive the formula (2.7) from (2.9).

For every 2×1 vector $\boldsymbol{a}$, it follows that $\mathrm{var}[Z]\geq 0$. Therefore, $\mathrm{var}[Z]=\boldsymbol{a}^T\boldsymbol{\Psi}_2\boldsymbol{a}\geq 0$. But, for any integer k such that $k\geq 2$ and $k\times 1$ vector $\boldsymbol{a}$ of real numbers, it can be seen that

$$\mathrm{var}[Z] = \boldsymbol{a}^T\boldsymbol{\Psi}_k\boldsymbol{a}\geq 0 \tag{2.10}$$

for all $k\times 1$ vectors $\boldsymbol{a}$ of real numbers. According to widely used terminology in matrix theory, any matrix Ψ_k is said to be semipositive definite if the equation (2.10) holds for all $k\times 1$ vectors $\boldsymbol{a}$ of real numbers. There is another case that is important when one is considering covariance matrices. Suppose that

$$\mathrm{var}[Z] = \boldsymbol{a}^T\boldsymbol{\Psi}_k\boldsymbol{a} > 0 \tag{2.11}$$

for all real vectors $\boldsymbol{a}$ such that $\boldsymbol{a}\neq\boldsymbol{0}$, a $k\times 1$ vector of zeroes. In this case the covariance matrix $\boldsymbol{\Psi}_k$ is said to be positive definite. As will be shown in what follows, there is an important connection between the covariance matrix being positive definite and determining whether an estimated covariance matrix for the case $k=2$ has the property such that ρ, the correlation coefficient, is guaranteed to belong to the open interval $(-1,1)$, as it should.

As discussed in Chapter 4, there are a number of criteria for determining whether a covariance matrix is positive definite for any $k\geq 2$. For the case

$k = 2$, it is easy to use a determinantal criterion to determine whether a covariance matrix is positive definite. Let

$$\Psi_2 = \begin{bmatrix} \sigma_1^2 & \rho\sigma_1\sigma_2 \\ \rho\sigma_2\sigma_1 & \sigma_2^2 \end{bmatrix} \tag{2.12}$$

denote a covariance matrix such that $\sigma_1^2 > 0$ and $\sigma_2^2 > 0$. Let $D_1 = \sigma_1^2$ and let

$$D_2 = \det \Psi_2 \tag{2.13}$$

be the determinate of Ψ_2. Then Ψ_2 is positive definite, if $D_1 > 0$ and $D_2 > 0$. By definition, $D_1 = \sigma_1^2 > 0$ by assumption. Furthermore, D_2 must satisfy the condition

$$D_2 = \sigma_1^2\sigma_2^2 - \rho^2\sigma_1^2\sigma_2^2 > 0. \tag{2.14}$$

But

$$D_2 = \sigma_1^2\sigma_2^2(1 - \rho^2) > 0. \tag{2.15}$$

Consequently, it follows that

$$1 - \rho^2 > 0, \tag{2.16}$$

which implies that $\rho^2 < 1$ and $\rho \in (-1, 1)$. When Ψ_2 is an estimate of a covariance, an investigator should apply the above test to determine whether the estimated matrix Ψ_2 is positive definite, which will ensure that the estimated ρ satisfies the condition $\rho \in (-1, 1)$. When the number k of rows and columns of a covariance matrix is $k > 2$, it is suggested that the reader consult Chapter 4, where several criteria for determining whether a covariance matrix is positive definite are discussed.

Although covariance matrices were not used formally in Mode and Robinson (1959), the procedures employed to estimate the unknown parameters could have been represented in terms of differences of covariance matrices. To illustrate this point, let A and B denote two 2×2 covariance matrices that would be part of a representation of an analysis of variance and covariance tables in matrix form. Suppose that an estimation procedure involves taking the difference of two covariance matrices. Then a matrix C of the form

$$C = A - B \tag{2.17}$$

arises. If it turns out that when the above test is carried out the matrix C is found to be positive definite, then one could proceed to estimate the parameter ρ such that $\rho \in (-1, 1)$. But, in general, there is no guarantee that C is positive definite. Consequently, an investigator may get estimates of ρ such that $\rho \notin (-1, 1)$. To avoid making such errors, the investigator should always test whether the matrix C is positive definite.

3. A One-Way Classification for Estimating Variance Components of a Quantitative Trait

When an investigator is dealing with domestic animals and plants and is investigating the quantitative genetics of one or more quantitative traits, it is possible to control the matings of individuals to produce relatives such that one can design experiments such that it is possible to estimate covariances among various types of relatives such as full sibs. By definition, relatives are said to be full sibs if they share two parents. But it is beyond the scope of this chapter to derive covariance matrices among relatives when one or more quantitative traits are under consideration. Consequently, in the types of models considered in what follows, it will be assumed that the sets of individuals under consideration are not related. The first statistical design that will be considered is known as a balanced one-way classification. It will be helpful to work out the formal details of this design, because the design used in the paper by Mode and Robinson (1959) may be viewed as an extension of the one-way classification.

For the case of one trait, let

$$Y_{ij} = \mu + \alpha_i + \epsilon_{ij} \tag{3.1}$$

denote a quantitative observation on an individual in the ith set of individuals, where $i = 1, 2, \ldots, a$ and a is a positive integer such that $a > 2$. The random variable α_i is a measure of the genetic effect in the ith set of individuals, and ϵ_{ij} is a random variable associated with the environmental effects within the ith set of individuals, where $j = 1, 2, \ldots, b$ for every individual $i = 1, 2, \ldots, a$. Furthermore, it is assumed that $E[\alpha_i] = 0$ for $i = 1, 2, \ldots, a$, where the expectation is with respect to some unknown genotypic distribution, and $E[\epsilon_{ij}] = 0$ for all pairs (i, j). Given these assumptions, it follows that

$$E[Y_{ij}] = \mu, \tag{3.2}$$

for $i = 1, 2, \ldots, a$ and $j = 1, 2, \ldots, b$.

It will also be assumed that all covariances $E[\alpha_i \alpha_{i'}] = 0$ for $i \neq i'$, $E[\epsilon_{ij}\epsilon_{ij'}] = 0$ for $j \neq j'$ and $E[\alpha_i \epsilon_{ij}] = 0$ for all pairs (i, j). Moreover, by definition, $E[\alpha_i^2] = \sigma_\alpha^2$ and $E[\epsilon_{ij}^2] = \sigma_\epsilon^2$ for each i and all pairs (i, j). From these assumptions, it follows that

$$\text{var}[Y_{ij}] = E[(Y_{ij} - \mu)^2] = \sigma_\alpha^2 + \sigma_\epsilon^2 \tag{3.3}$$

for all pairs (i, j). Thus, by assumption, all observations have a common expectation and variance.

It is of fundamental interest to find estimates of the terms on the right-hand side of (3.1) based on the data. An estimate of the expectation μ is the sample mean, which is given by the formula

$$\overline{Y} = \frac{1}{ab} \sum_{i=1}^{a} \sum_{j=1}^{b} Y_{ij}. \tag{3.4}$$

Another mean that is of interest is

$$\overline{Y}_i = \frac{1}{b} \sum_{j=1}^{b} Y_{ij}, \tag{3.5}$$

for every $i = 1, 2, \ldots, a$. Observe that $\overline{Y}_i$ is the mean of the measurements on the b individuals in set i.

Next, consider the identity

$$Y_{ij} = \overline{Y} + (\overline{Y}_i - \overline{Y}) + (Y_{ij} - \overline{Y}_i), \tag{3.6}$$

which is valid for all pairs (i, j), and observe that

$$Y_{ij} - \overline{Y} = (\overline{Y}_i - \overline{Y}) + (Y_{ij} - \overline{Y}_i). \tag{3.7}$$

For every fixed $i = 1, 2, \ldots, a$, square both sides of the equation (3.7) to obtain

$$(Y_{ij} - \overline{Y})^2 = (\overline{Y}_i - \overline{Y})^2 + (Y_{ij} - \overline{Y})^2 + 2(\overline{Y}_i - \overline{Y})(Y_{ij} - \overline{Y}_i). \tag{3.8}$$

Then, for every fixed i, sum the equation (3.8) over j to obtain

$$\sum_{j=1}^{b} (Y_{ij} - \overline{Y})^2 = b(\overline{Y}_i - \overline{Y})^2 + \sum_{j=1}^{b} (Y_{ij} - \overline{Y}_i)^2$$

$$+ 2(\overline{Y}_i - \overline{Y}) \sum_{j=1}^{b} (Y_{ij} - \overline{Y}_i). \tag{3.9}$$

But

$$\sum_{j=1}^{b}(Y_{ij}-\overline{Y}_i)=\sum_{j=1}^{b}Y_{ij}-b\overline{Y}_i=\sum_{j=1}^{b}Y_{ij}-\sum_{j=1}^{b}Y_{ij}=0, \tag{3.10}$$

for every $i = 1, 2, \ldots, a$. Note that this equation follows from the definition of $\overline{Y}_i$ in the equation (3.5). Thus,

$$\sum_{j=1}^{b}(Y_{ij}-\overline{Y}_i)^2=b(\overline{Y}_i-\overline{Y})^2+\sum_{j=1}^{b}(Y_{ij}-\overline{Y}_i)^2, \tag{3.11}$$

for every i. By summing over i it can be seen that

$$\sum_{i=1}^{a}\sum_{j=1}^{b}(Y_{ij}-\overline{Y})^2=b\sum_{i=1}^{a}(\overline{Y}_i-\overline{Y})^2+\sum_{i=1}^{a}\sum_{j=1}^{b}(Y_{ij}-\overline{Y}_i)^2. \tag{3.12}$$

This is a partition of the total sum of squares about the mean of the data into two components on the right. The first term on the right is a measure of the variation of the means for the sets of individuals about the sample mean $\widehat{Y}$ in (3.4), and the second term is a measure of the variation in the sets of individuals around the mean of each set. It can be shown by using the equation (3.1) that

$$E\left[b\sum_{i=1}^{a}(\overline{Y}_i-\overline{Y})^2\right]=(a-1)(\sigma_\epsilon^2+b\sigma_\alpha^2), \tag{3.13}$$

$$E\left[\sum_{i=1}^{a}\sum_{j=1}^{b}(Y_{ij}-\overline{Y}_i)^2\right]=a(b-1)\sigma_\epsilon^2. \tag{3.14}$$

Thus,

$$\frac{1}{a-1}E\left[\sum_{i=1}^{a}\sum_{j=1}^{b}(Y_{ij}-\overline{Y}_i)^2\right]=\sigma_\epsilon^2+b\sigma_\alpha^2, \tag{3.15}$$

$$\frac{1}{a(b-1)}E\left[\sum_{i=1}^{a}\sum_{j=1}^{b}(Y_{ij}-\overline{Y}_i)^2\right]=\sigma_\epsilon^2. \tag{3.16}$$

Therefore,

$$\frac{1}{a(b-1)}\sum_{i=1}^{a}\sum_{j=1}^{b}(Y_{ij}-\overline{Y}_i)^2 = q_2 \tag{3.17}$$

is an estimator of σ_ϵ^2 and

$$\frac{1}{a-1}\sum_{i=1}^{a}\sum_{j=1}^{b}(Y_{ij}-\overline{Y}_i)^2 = q_1 \tag{3.18}$$

is an estimator of $\sigma_\epsilon^2 + b\sigma_\alpha^2$.

From these equations, it can be seen that

$$q_1 - q_2 = b\sigma_\alpha^2, \tag{3.19}$$

so that

$$\widehat{\sigma_\alpha^2} = \frac{q_1 - q_2}{b} \tag{3.20}$$

is an estimator of σ_α^2, given the data. If $q_1 > q_2$, then $\widehat{\sigma_\alpha^2} > 0$. But if $q_1 \leq q_2$, then $\widehat{\sigma_\alpha^2} \leq 0$. It is of interest to note that

$$E[\widehat{\sigma_\alpha^2}] = E\left[\frac{q_1 - q_2}{b}\right] = \frac{1}{b}(E[q_1 - q_2]) = \frac{1}{b}(\sigma_\epsilon^2 + b\sigma_\alpha^2) - \frac{\sigma_\epsilon^2}{b} = \sigma_\alpha^2, \tag{3.21}$$

which shows that the expression (3.20) is an unbiased estimator of σ_α^2. It is interesting to note that even though $\widehat{\sigma_\alpha^2}$ may be negative in some samples, it is unbiased with respect to all possible samples of data with ab observations. In the appendix, proofs of the formulas (3.13) and (3.14) are given.

4. A Multivariate Version of the One-Way Classification for Estimating Variance and Covariance Components of Multiple Traits

In this section a multivariate version of the one-way classification described in Section 3 will be generalized to the case in which two or more quantitative traits can be accommodated. As in Section 3, the case of $a \geq 2$ classifications will be considered, and suppose that $b \geq 2$ observations for each class have been made. It will also be assumed that each observation is vector-valued with $c \geq 1$ elements. In terms of quantitative genetics, $c \geq 1$ traits are under consideration. In class i, let Y_{ijk} denote observation j on trait k, for $i = 1, 2, \ldots, a$, $j = 1, 2, \ldots, b$ and $k = 1, 2, \ldots, c$. It will be assumed that

$$Y_{ijk} = \mu_k + \alpha_{ik} + \epsilon_{ijk}. \tag{4.1}$$

For a fixed pair (i, j), let

$$\boldsymbol{Y}_{ij} = \boldsymbol{\mu} + \boldsymbol{\alpha}_i + \boldsymbol{\epsilon}_{ij}, \tag{4.2}$$

where $\boldsymbol{\mu}, \boldsymbol{\alpha}_i$ and $\boldsymbol{\epsilon}_{ij}$ are $c \times 1$ vectors. In terms of their transposes,

$$\begin{aligned}
\boldsymbol{\mu}^T &= (\mu_1, \mu_2, \ldots, \mu_c), \\
\boldsymbol{\alpha}_i^T &= (\alpha_{i1}, \alpha_{i2}, \ldots, \alpha_{ic}), \\
\boldsymbol{\epsilon}_{ij}^T &= (\epsilon_{ij1}, \epsilon_{ij2}, \ldots, \epsilon_{ijc}).
\end{aligned} \tag{4.3}$$

It will also be supposed that $E[\boldsymbol{\alpha}_j] = \boldsymbol{0}$ and $E[\boldsymbol{\epsilon}_{ij}] = \boldsymbol{0}$, where $\boldsymbol{0}$ is a $c \times 1$ vector of zeroes. In addition, it will be assumed that

$$E\left[\boldsymbol{\alpha}_i \boldsymbol{\alpha}_{i'}^T\right] = \boldsymbol{0}_{c \times c} \tag{4.4}$$

if $i \neq i'$, where $\boldsymbol{0}_{c \times c}$ is a square matrix of zeroes with c rows and columns. Similarly, it will be assumed that

$$E\left[\boldsymbol{\epsilon}_{ij} \boldsymbol{\epsilon}_{ij'}^T\right] = \boldsymbol{0}_{c \times c} \tag{4.5}$$

if $j \neq j'$ and

$$E\left[\boldsymbol{\epsilon}_{ij} \boldsymbol{\alpha}_i^T\right] = \boldsymbol{0}_{c \times c} \tag{4.6}$$

for all pairs (i, j).

The next step in formulating the model is to define the following covariance matrices. For all pairs (i, j), the following definitions of $c \times c$ covariance matrices will play a fundamental role in the formulation of a multivariate one-way classification. By definition,

$$E\left[(\boldsymbol{Y}_{ij} - \boldsymbol{\mu})(\boldsymbol{Y}_{ij} - \boldsymbol{\mu})^T\right] = \boldsymbol{\Psi}, \tag{4.7}$$

$$E\left[\boldsymbol{\alpha}_{ij} \boldsymbol{\alpha}_{ij}^T = \boldsymbol{\Psi}_\alpha\right], \tag{4.8}$$

$$E\left[\boldsymbol{\epsilon}_{ij} \boldsymbol{\epsilon}_{ij}^T\right] = \boldsymbol{\Psi}_\epsilon. \tag{4.9}$$

As a first step in proving the validity of the matrix equation

$$\boldsymbol{\Psi} = \boldsymbol{\Psi}_\alpha + \boldsymbol{\Psi}_\epsilon, \tag{4.10}$$

consider

$$E[(Y_{ij} - \mu)(Y_{ij} - \mu)^T] = E[(\boldsymbol{\alpha}_i + \boldsymbol{\epsilon}_{ij})(\boldsymbol{\alpha}_i + \boldsymbol{\epsilon}_{ij})^T]. \tag{4.11}$$

But

$$(\boldsymbol{\alpha}_i + \boldsymbol{\epsilon}_{ij})(\boldsymbol{\alpha}_i + \boldsymbol{\epsilon}_{ij})^T = (\boldsymbol{\alpha}_i + \boldsymbol{\epsilon}_{ij})\boldsymbol{\alpha}_i^T + (\boldsymbol{\alpha}_i + \boldsymbol{\epsilon}_{ij})\boldsymbol{\epsilon}_{ij}^T. \tag{4.12}$$

Next, observe that

$$E\big[(\boldsymbol{\alpha}_i + \boldsymbol{\epsilon}_{ij})\boldsymbol{\alpha}_i^T\big] = E\big[\boldsymbol{\alpha}_i\boldsymbol{\alpha}_i^T\big] + E\big[\boldsymbol{\epsilon}_{ij}\boldsymbol{\alpha}_i^T\big]. \tag{4.13}$$

By assumption,

$$E\big[\boldsymbol{\epsilon}_{ij}\boldsymbol{\alpha}_i^T\big] = \mathbf{0}_{c\times c}. \tag{4.14}$$

Therefore,

$$E\big[(\boldsymbol{\alpha}_i + \boldsymbol{\epsilon}_{ij})\boldsymbol{\alpha}_i^T\big] = E\big[\boldsymbol{\alpha}_i\boldsymbol{\alpha}_i^T\big] = \boldsymbol{\Psi}_\alpha. \tag{4.15}$$

Similarly, it can be shown that

$$E\big[(\boldsymbol{\alpha}_i + \boldsymbol{\epsilon}_{ij})\boldsymbol{\epsilon}_{ij}^T\big] = E\big[\boldsymbol{\epsilon}_{ij}\boldsymbol{\epsilon}_{ij}^T\big] = \boldsymbol{\Psi}_\epsilon. \tag{4.16}$$

It may be concluded, therefore, that the equation (4.10) is valid, given the assumptions stated above.

The next step in setting up a structure such that the covariance matrices may be estimated is to find estimates of all vectors on the right-hand side of (4.2), using the data. An estimate of the expectation vector $\boldsymbol{\mu}$ is the sample mean vector

$$\overline{Y} = \frac{1}{ab}\sum_{i=1}^{a}\sum_{j=1}^{b} Y_{ij}. \tag{4.17}$$

The estimation of other mean vectors will also be necessary to find estimates of the other vectors on the right-hand side of (4.2). For a fixed i, consider the vector

$$\overline{Y}_i = \frac{1}{b}\sum_{j=1}^{b} Y_{ij}. \tag{4.18}$$

Then observe that

$$Y_{ij} = \overline{Y} + (\overline{Y}_i - \overline{Y}) + (Y_{ij} - \overline{Y}_i) \tag{4.19}$$

is a valid equation for all pairs (i, j). In order to set up an analysis of covariance procedure, to find an estimator of the matrices involved in the estimation procedure, let

$$Q = \sum_{i=1}^{a}\sum_{j=1}^{b}(Y_{ij} - \overline{Y})(Y_{ij} - \overline{Y})^T \tag{4.20}$$

denote a $c \times c$ matrix. Next, let Q_1 denote the $c \times c$ matrix

$$Q_1 = \sum_{i=1}^{a}(\overline{Y}_i - \overline{Y})(\overline{Y}_i - \overline{Y})^T. \tag{4.21}$$

Finally, let Q_2 denote the $c \times c$ matrix

$$Q_2 = \sum_{i=1}^{a}\sum_{j=1}^{b}(Y_{ij} - \overline{Y}_i)(Y_{ij} - \overline{Y}_i)^T. \tag{4.22}$$

Then it can be shown that

$$Q = Q_1 + Q_1. \tag{4.23}$$

The validity of this equation may be shown by using the procedure set forth in (3.9) and (3.10) for each element of the matrices Q_1 and Q_2. It should also be observed that the technique applied in (3.9) and (3.10) for one trait also applies to each sum of product terms that are located on the principal diagonal on the matrices in (4.23). Let

$$m_1 = \frac{1}{a-1}Q_1, \tag{4.24}$$

$$m_2 = \frac{1}{a(b-1)}Q_2. \tag{4.25}$$

Then it can be shown that

$$E[m_1] = \mathbf{\Psi}_\epsilon + b\mathbf{\Psi}_\alpha. \tag{4.26}$$

Similarly, it can be shown that

$$E[m_2] = \mathbf{\Psi}_\epsilon. \tag{4.27}$$

To verify the validity of these two equations, it is suggested that the formulas presented in the outline in the appendix for the case of one trait be extended to two or more traits.

Therefore, an estimator of the matrix $\mathbf{\Psi}_\alpha$ is given by the formula

$$\widehat{\mathbf{\Psi}}_\alpha = \frac{1}{b}(m_1 - m_2). \tag{4.28}$$

It is recommended that the estimate $\widehat{\mathbf{\Psi}}_\alpha$ should be tested as to whether it is positive definite. For the case $c = 2$, this test would be straightforward, but for cases such that $c > 2$, the procedures for testing whether a matrix is positive definite described in Chapter 5 should be consulted. It is interesting to note that

$$E[\widehat{\mathbf{\Psi}}_\alpha] = \frac{1}{b}(\mathbf{\Psi}_\epsilon + b\mathbf{\Psi}_\alpha - \mathbf{\Psi}_\epsilon) = \mathbf{\Psi}_\alpha, \tag{4.29}$$

so that $\widehat{\mathbf{\Psi}}_\alpha$ is an unbiased estimator of $\mathbf{\Psi}_\alpha$.

5. An Experimental Design and Data

Design I of Comstock and Robinson is a generalization of the structures considered in Sections 4 and 5. The experimental material was produced from matings among plants of the F_2 generation of a cross between inbred lines of field corn (Zia mays). Like many plants in the grass family, field corn is a hermaphrodite. The male organ is the tassel, because it is in this organ that the pollen is produced. The ear of the plant is the female organ, because it contains the eggs that may be fertilized by pollen from the tassel. During pollination time, structures called silk strands appear on the ear and are catchers of the pollen. If pollen falls on a silk strand, then by a remarkable process it goes to the eggs in an ear and fertilizes them to produce the kernels attached to the cob of each ear. Mating may be controlled by putting paper bags on the tassel and ears of a plant or plants. In a particular mating the male furnishes the pollen which is placed on the silk of ears on the female plants. As mentioned above, the experimental material labeled males and females was produced among plants of the F_2 generation of a mating to two inbred lines of corn.

A random sample sm of plants from the F_2 generation was chosen and mated to a random sample of n females. No male was used more than once. In particular, the experimental material is of the offspring of these mn matings and is partitioned into s sets. It is well known that the quality of the soil may vary within short distances in an experimental field. The planting of some number of sets within a field takes into account variability in the quality of the soils in a field. Each set thus forms a distinct unit of the experiment and is planted in a classical randomized block design, with mn entries and r replications. The offspring in each set are those from the m males mated to n females in a set. Quantitative measurements were on about 10 plants in a plot and the plot means were recorded for each trait.

For the case of two traits, the mean of two traits may be represented by the vector-valued model

$$Y_{ijkl} = \boldsymbol{\mu} + \boldsymbol{s}_i + \boldsymbol{r}_{ij} + \boldsymbol{m}_{ik} + \boldsymbol{f}_{ikl} + \boldsymbol{\varepsilon}_{ijkl}, \tag{5.1}$$

for $i = 1, 2, \ldots, s, j = 1, 2, \ldots, r, k = 1, 2, \ldots, m$ and $l = 1, 2, \ldots, n$. Each vector in (5.1) has two rows and one column. In terms of their transposes, the parameters of the model are $\boldsymbol{\mu}^T = (\mu_1, \mu_2)$ and the population means for each trait, $\boldsymbol{s}_i^T = (s_{1i}, s_{2i})$ are the effects for the ith set for the two traits. Similarly, $\boldsymbol{r}_{ij}^T = (r_{ij}, r_{2ij})$, are the effects of the two traits in the jth replication of the ith set, and $\boldsymbol{m}_{ik}^T = (m_{1ik}, m_{2ik})$ are the effects for of the kth male of the ith set. Finally, $\boldsymbol{f}_{ikl}^T = (f_{1ikl}, f_{2ikl})$ are the effects for the lth female mated to the kth male in the ith set, and the vector $\boldsymbol{\epsilon}_{ijkl} = (\epsilon_{1ijkl}, \epsilon_{2ijkl})$ means random errors. Moreover, it will be assumed that the vectors $\boldsymbol{m}_{ik}, \boldsymbol{f}_{ikl}$ and $\boldsymbol{\varepsilon}_{ijkl}$ are independent bivariate normal random variables with expectations $\mathbf{0}$, a two-by-one vector of zeroes, with the covariance matrices

$$\boldsymbol{\Psi}_m = \begin{bmatrix} \sigma_{1m}^2 & \sigma_{12m} \\ \sigma_{21m} & \sigma_{2m}^2 \end{bmatrix}, \tag{5.2}$$

$$\boldsymbol{\Psi}_f = \begin{bmatrix} \sigma_{1f}^2 & \sigma_{f12} \\ \sigma_{21f} & \sigma_{2f}^2 \end{bmatrix}, \tag{5.3}$$

$$\boldsymbol{\Psi}_\epsilon = \begin{bmatrix} \sigma_{1\epsilon}^2 & \sigma_{\epsilon 12} \\ \sigma_{\epsilon 21} & \sigma_{\epsilon 2}^2 \end{bmatrix}. \tag{5.4}$$

It will also be assumed that the vectors $\boldsymbol{s}_i$ and $\boldsymbol{r}_{ij}$ are fixed so that

$$\sum_{i=1}^{s} \boldsymbol{s}_i = \boldsymbol{0} \tag{5.5}$$

and

$$\sum_{i=1}^{r} \boldsymbol{r}_{ij} = \boldsymbol{0}, \tag{5.6}$$

for $j = 1, 2, \ldots, r$. From the assumptions that the vectors $\boldsymbol{m}_{ik}, \boldsymbol{f}_{ikl}$ and $\boldsymbol{\varepsilon}_{ijkl}$ are independent with zero mean vectors and covariance matrices (5.2), (5.3) and (5.4), it follows that estimates of their covariances matrices have independent Wishart distributions; see Anderson (1958) and subsequent editions of the book.

In Table 1.1 a theoretical analysis of variance covariance analysis of a set of data is presented.

In the third column of the table, each symbol stands for an estimated covariance matrix based on the set of data. Procedures for computing the elements of these matrices from the data have been listed in Mode and Robinson (1959) for the case of two traits.

In the analysis of data to be presented in what follows, the two traits under consideration were plant height and ear height. The data consisted of

Table 1.1. Theoretical Covariance Analysis

Source of Variation	Degrees of Freedom	Matrix
Sets	$s-1$	S
Replications	$s(r-1)$	R
Males	$s(m-1)$	M
Females	$sm(n-1)$	F
Remainder	$s(mn-1)(r-1)$	E

Table 1.2. Covariance Analysis for Plant and Ear Height

Source of Variation	Degrees of Freedom	Mean Matrices
Sets	15	m_S
Replications	16	m_R
Males	48	m_M
Females	192	m_F
Remainder	239	m_{RM}

observations on progenies of plants that were observed in accordance with Design I of Comstock and Robinson. The two traits under consideration were measured on plants that were of offspring of plants in the F_2 generation from a cross of two inbred lines, $C121$ and $NC7$. In this case, $s = 16, m = 4, n = 4$ and $r = 2$. By definition, the mean square matrix for matrix S is $m_S = S/(s-1)$. Similarly, all other mean square matrices were computed by dividing each matrix by the corresponding degrees of freedom. In Table 1.2 the mean squares that were estimated from the data are presented in terms of their elements. The summarized data from this experiment are shown in the table.

The numerical elements of the mean matrices are as follows:

$$m_S = \begin{bmatrix} 235.9540 & 84.5967 \\ 84.5967 & 53.8533 \end{bmatrix}, \tag{5.7}$$

$$m_R = \begin{bmatrix} 28.2498 & 14.7096 \\ 14.7096 & 12.55226 \end{bmatrix}, \tag{5.8}$$

$$m_M = \begin{bmatrix} 77.6481 & 35.5160 \\ 35.5160 & 38.6442 \end{bmatrix}, \tag{5.9}$$

$$m_F = \begin{bmatrix} 30.6799 & 15.0266 \\ 15.0266 & 11.4386 \end{bmatrix}, \tag{5.10}$$

$$m_{RM} = \begin{bmatrix} 10.0099 & 4.1128 \\ 4.1128 & 4.0039 \end{bmatrix}. \tag{5.11}$$

These matrices estimated from the data are matrix-valued random variables with the expectations for the matrices m_M, m_F and m_{RM}. It can be shown that

$$E[m_{RM}] = \boldsymbol{\Psi}_\epsilon, \tag{5.12}$$

$$E[m_F] = \boldsymbol{\Psi}_\epsilon + r\boldsymbol{\Psi}_f, \tag{5.13}$$

$$E[m_M] = \boldsymbol{\Psi}_\epsilon + r\boldsymbol{\Psi}_f + rn\boldsymbol{\Psi}_m. \tag{5.14}$$

By the method of moments for matrices, it can be seen that

$$\widehat{\boldsymbol{\Psi}}_\epsilon = m_{RM} \tag{5.15}$$

is an estimator of the matrix $\boldsymbol{\Psi}_\epsilon$. From the equation (5.14), it can be seen that

$$\widehat{\boldsymbol{\Psi}}_f = \frac{1}{r}(m_F - m_{RM}) \tag{5.16}$$

is an estimator of the matrix $\boldsymbol{\Psi}_f$. Similarly, it can be seen that

$$\widehat{\boldsymbol{\Psi}}_m = \frac{1}{rn}(m_M - m_F) \tag{5.17}$$

is an estimator of the matrix $\boldsymbol{\Psi}_m$. It will be left as an exercise for the reader to show that all these matrix-valued estimators are unbiased. Because the random matrices m_M, m_F and m_{RM} have independent Wishart distributions, the moment-generating functions of these distributions can be used to estimate standard errors of these estimates, as was done in Mode and Robinson (1959).

After finding an estimator of a covariance matrix, the next step in the estimation process is to determine whether the estimator has the necessary properties of a covariance matrix. In this connection, the matrix m_{RM} will be tested as to whether it is positive definite. From (5.11) it can be seen that $m_{RM}(1,1) = 10.0099 > 0$. It can also be shown that the determinant of the matrix is 23.164. Therefore, it can be concluded that the matrix estimator in (5.15) is positive definite, and it follows that the estimate correlation coefficient ρ_ϵ satisfies the condition $-1 < \rho_\epsilon < 1$, as it should. From now on, to determine whether a covariance is positive definite, it will suffice to compute the determinate of the matrix, because all the elements in the matrices that follow are positive.

With regard to testing the matrix $\widehat{\boldsymbol{\Psi}}_f$ in (5.16) as to whether it is positive definite, it can be shown that

$$m_F - m_{RM} = \begin{bmatrix} 20.67 & 10.914 \\ 10.914 & 7.4347 \end{bmatrix}. \tag{5.18}$$

The determinate of this matrix is 34.560. Thus, the matrix in (5.16) has the property of being positive definite, as it should. Similarly, to test whether the matrix estimator in (5.17) is positive definite, it can be shown that

$$m_M - m_F = \begin{bmatrix} 46.968 & 20.489 \\ 20.489 & 27.206 \end{bmatrix}, \tag{5.19}$$

and the determinate of this matrix is 858.01. Because $rn > 0$, it may be concluded that the estimator in (5.19) is positive definite.

Let σ_{R1}^2 and σ_{R2}^2 denote variance components of the two traits with respect to replications in sets in the model under consideration. From the summarized data from the experiment under consideration as displayed in (5.9) and (5.10), it can be seen that the elements of the matrix

$$m_R - m_M \tag{5.20}$$

are all negative. Consequently, it is not possible to obtain a positive estimates of the parameters σ_{R1}^2 and σ_{R2}^2 based on the data under consideration. Fortunately, these parameters were of no interest from the genetic point of view in the data that were analyzed. However, this in an example that the class of experimental designs under consideration has the flaw in that unacceptable estimates of covariance matrices may be observed in some sets of data. All investigators who use such experimental designs should be aware of this potential flaw. Furthermore, it will be a theoretical challenge for theoreticians to design methods of estimation that will avoid such flaws.

In the introduction, the idea of additive and dominant covariance matrices was introduced. Let

$$\boldsymbol{\Psi}_A = \begin{bmatrix} \sigma_{A1}^2 & \sigma_{A12} \\ \sigma_{A21} & \sigma_{A2}^2 \end{bmatrix} \tag{5.21}$$

denote the additive covariance matrix, and let

$$\boldsymbol{\Psi}_D = \begin{bmatrix} \sigma_{D1}^2 & \sigma_{D12} \\ \sigma_{D21} & \sigma_{D2}^2 \end{bmatrix} \tag{5.22}$$

denote the dominance covariance matrix for the two traits under consideration. The idea of a genetic covariance matrix was also introduced in the introduction. Therefore, let

$$\boldsymbol{\Psi}_G = \begin{bmatrix} \sigma_{G1}^2 & \sigma_{G12} \\ \sigma_{G21} & \sigma_{G2}^2 \end{bmatrix} \tag{5.23}$$

denote the genetic covariance matrix. Because the plants labeled female in the experiment were the offspring of random matings, it will be assumed that the experimental population was in a Hardy–Weinberg equilibrium. Under these conditions, it can be shown that

$$\boldsymbol{\Psi}_G = \frac{1}{2}\boldsymbol{\Psi}_A + \frac{1}{4}\boldsymbol{\Psi}_D. \tag{5.24}$$

This formula follows from the assumptions underlying the experiment, because the frequency of the allele is $1/2$ and the mating is random.

To meet the goals of the experiment stated in the introduction, it will be necessary to obtain estimates of all the matrices in the equation (5.24). Because in Design 1 of Comstock and Robinson (1948), the plants labeled female in the experiment were the offspring of random matings, it follows that the matrix in (5.16) is an estimator of the matrix $\boldsymbol{\Psi}_G$. Thus,

$$\widehat{\boldsymbol{\Psi}}_G = \widehat{\boldsymbol{\Psi}}_F = \frac{1}{2}\widehat{\boldsymbol{\Psi}}_A + \frac{1}{4}\widehat{\boldsymbol{\Psi}}_D. \tag{5.25}$$

The experiment was also designed in such a way that the plants labeled male were half sibs, because such plants had a common male parent. For the case of one trait, the covariance among parent and offspring has the special form

$$\operatorname{cov}[P, O] = \frac{1}{2}\sigma_A^2. \tag{5.26}$$

For the case of two or more traits, it can also be shown that

$$\operatorname{cov}[P, O] = \frac{1}{2}\boldsymbol{\Psi}_A. \tag{5.27}$$

Let $\widehat{\boldsymbol{\Psi}}_A$ denote an estimator of the covariance matrix $\boldsymbol{\Psi}_A$. Then, from the equation (5.27) and the assumptions underlying the experiment, it follows that

$$\frac{1}{2}\widehat{\mathbf{\Psi}}_A = \widehat{\mathbf{\Psi}}_m, \tag{5.28}$$

so that

$$\widehat{\mathbf{\Psi}}_A = 2\widehat{\mathbf{\Psi}}_m \tag{5.29}$$

is an estimate of the matrix $\mathbf{\Psi}_A$, and

$$\widehat{\mathbf{\Psi}}_D = 4(\widehat{\mathbf{\Psi}}_f - \widehat{\mathbf{\Psi}}_m) \tag{5.30}$$

is an estimate of the covariance matrix $\mathbf{\Psi}_D$. In order to apply the formulas (5.29) and (5.30) using the data, it will be necessary to find the numerical form of the matrices listed in (5.16) and (5.17). Given the formula (5.17) and the data, it can be shown that

$$\widehat{\mathbf{\Psi}}_m = \begin{bmatrix} 5.871 & 2.5611 \\ 2.5611 & 3.4008 \end{bmatrix}. \tag{5.31}$$

Therefore, it follows from the equation (5.29) that

$$\widehat{\mathbf{\Psi}}_A = \begin{bmatrix} 11.742 & 5.1222 \\ 5.1222 & 6.8016 \end{bmatrix} \tag{5.32}$$

is the estimate of the additive covariance matrix. In summary, it can be shown that

$$\widehat{\mathbf{\Psi}}_f = \begin{bmatrix} 15.340 & 7.5133 \\ 7.5133 & 5.7193 \end{bmatrix}. \tag{5.33}$$

Thus, by applying the formula (5.30) it can be seen that

$$\widehat{\mathbf{\Psi}}_D = \begin{bmatrix} 37.876 & 19.809 \\ 19.809 & 9.274 \end{bmatrix} \tag{5.34}$$

is an estimate of the dominance covariance matrix.

It is interesting to note that all the elements of the estimated dominance covariance matrix in (5.34) are uniformly greater than the elements of the additive covariance matrix in (5.32). It is also of interest to check whether the matrices in (5.32) and (5.34) are positive definite. From (5.32), it can be seen that all the elements of the additive covariance matrix are positive, and it can be shown that the determinant of the matrix is 53.627. Consequently, the additive covariance matrix is positive definite. However, even though all the elements of the dominant covariance matrix in (5.34) are positive, it can be shown that its determinant is −41.134. Therefore, the dominant covariance matrix is not positive definite. Interestingly, this is an example that the difference of two positive definite matrices may not be positive definite, as was suggested in the introduction. In this connection, it is of interest to compute the correlation coefficient of the dominant covariance matrix. It can be shown that

$$\rho_D = \frac{19.809}{\sqrt{37.876 \times 9.274}} = 1.0569. \tag{5.35}$$

This estimate lies outside of the interval $(-1, 1)$, and is therefore cause for concern. However, from the practical point of view, it seems safe to conclude that the dominant effects of plant and ear height are highly correlated and positive. This example based on real data points to a need to develop an estimation procedure for the experimental design under consideration as well as other designs that will produce estimates of covariance matrices that are positive definite or at least matrices that are nonnegative definite. It is interesting to note that even though the dominant covariance matrix was not positive definite, estimates of the average dominance were possible to obtain for both traits, plant and ear height, by using the formula (1.13), as was shown by Mode and Robinson (1959). Moreover, because the estimated additive matrix in (5.32) is positive definite and all its elements are positive, it follows that the additive correlation coefficient belongs to the interval $(-1, 1)$ and is positive, which was also shown by Mode and Robinson (1959).

From extensive studies of crosses of inbred lines of field corn, measures of quantitative traits of plants in the F_1 generation often exceed measures of the corresponding traits in the parental generation consisting of inbred lines. This phenomenon is sometimes called the hybrid effect, and it is also known as heterosis. The data analysis just described suggests that the hybrid effects may also be observed in populations resulting from crosses among plants in the F_2 generation. It is also interesting to observe that if only genes at one locus with two alleles govern the quantitative genetics of plant and ear height, then it follows from Mendel's laws that about half of individuals making up the population of offspring of random matings among individuals in an F_2 generation will have the same genotype as individuals in the F_1 population that result from crosses of two inbred lines of corn. The other half of the population will consist of individuals with the same genotype as those in the two inbred lines that were under consideration in an experiment. This observation suggests that if the F_1 generation exhibits a heterotic effect, then this effect will also account for the presence of this effect in a population that arises from random matings among individuals of the F_2 generation in which half of the individuals have the same genotype as individuals in the F_1 generation.

The illustrative example just outlined is of value, but it should be emphasized that in the 1950s, when the experiment was conducted, investigators did not know where the regions were located in the genome of a species that were involved in the expression of a quantitative trait. In the present age of genomics, however, many investigators have found regions in a genome that suggests there is evidence that they are involved in the expression of a quantitative trait. Given the tentative information as to where these regions are located in the genome of a species and working definitions of at least two alleles at each region (locus), the ideas just discussed for the case of one autosomal locus with two alleles can be extended to define various types of measures of epistasis, which were mentioned in the paper by Mode and Robinson (1959) but could not be estimated. In Chapters 3 and 4 it is shown that measure of epistasis can be defined and estimated, given working definitions of a set of loci with two or more alleles by a simple but very useful direct method. Moreover, as will be demonstrated in Chapter 4, there is no need to use the differences of two positive definite covariances to obtain estimates of epistatic effects.

6. A Prospective View of Mode and Robinson (1959)

As mentioned before, during the 1950s desktop computers as we know them today, along with the software to run them were inconceivable, but there were signs of coming developments, such as seminars that were presented on writing software in a very-low-level language that would work on computers that were available at that time, for example the IBM 650. My impression of these seminars was that they were interesting but it would be too tedious to pursue this line of work. Consequently, as a mathematical geneticist who was just starting his career, I decided to undertake an in-depth study of the mathematics underlying probability and statistics rather than pursuing what is now known as computer science. The statistical ideas applied in the 1959 paper were a sign of the level of mathematical statistics that I was comfortable using in 1958. But, as other investigators used this work, problems arose in the estimation of parameters, such as estimates of a variance that was negative rather than positive as it should be, or an estimate of a correlation coefficient that was outside the interval $(-1, 1)$. These problems lingered in my mind for many years but, because I devoted much of my research efforts to developing branching processes and their applications, these estimation problems that were encountered while I was working on quantitative genetics were neglected. It was not until the last two years, when I returned to the study of quantitative genetics, that solutions to these estimation problems were realized in terms of properties of covariance matrices, such as the matrix being positive definite.

At the time this paper was written in 1958, all calculations were done with desktop calculating machines, which made it necessary to enter all numbers used in a calculation procedure into the machine, through a keyboard. Such methods sometimes led to errors in the input of data so that disagreements arose among investigators who got different values when parameters were estimated. Another source of error was that at some points in a calculation procedure, a value that had been computed previously needed to be re-entered into the machine, which also led to errors in the final result. In the present computer age, errors in data input may also occur, but if the computer code is written correctly, then, given a sample of numbers, the computer output will be essentially correct to a number of decibels, depending on the word length of the computer. Given such potential problems, in 1958 the calculations involving matrices were not undertaken, because it would have been very tedious to check whether the results were correct. However, because a calculation engine was included as part of the word processor used to write this chapter, it was possible to do the matrix calculations shown in Section 5 with a high level of confidence that they were indeed correct.

It seems likely that experimental designs such as Design I of Comstock and Robinson, described in Section 5, are still being used today. Consequently, given the availability of software to do matrix calculations, it would be straightforward to write software that would automatically do the matrix calculations listed in Section 5 and check whether estimated covariance matrices were positive definite. But the potential problems that may happen when one is taking the difference of two covariance matrices that are positive definite may appear with some regularity. Consequently, methods of estimation will need to be developed such that estimated covariance matrices are at least positive semidefinite or positive definite. One approach to overcoming such problems is to use a Bayesian approach to estimating parameters. In this connection, it is interesting to note that, given the normality assumptions stated for the model for Design I of Comstock and Robinson, it would be possible to derive the likelihood function of the data.

Briefly, if a Bayesian approach were used, then a prior distribution would be chosen such that given the data, covariance matrices would be estimated from the posterior distribution using the formula for the conditional expectation

$$E[(X-\mu)(X-\mu)^T \mid D] = \boldsymbol{\Psi}, \tag{6.1}$$

where D is the data set under consideration. The use of this formula guarantees that estimates of covariance matrices would be at least positive semidefinite and in many cases would be positive definite.

When one is working within a Bayesian paradigm, a parameter or a set of parameters is estimated by computing a conditional expectation with respect to the posterior distribution. Quite often, difficult problems arise when one is trying to find a useful form of a conditional expectation such as that in (6.1), using classical formal integration techniques. A technique called Monte Carlo integration is often useful for finding a numerical value of a complex integral. For example, let $X_1, X_2, \ldots, X_N$ be a sample of independent Monte Carlo realizations of a sample from the posterior distribution. The simulated Monte Carlo sample may consist of a set of real numbers, for example, or a set of column vectors when an investigator is dealing with two or more quantitative traits. Suppose that the simulated Monte Carlo sample is a set of realized vectors when a model of several traits is under consideration. Then, by invoking the strong law of large numbers, it follows that

$$\lim_{N\uparrow\infty} \frac{1}{N}\sum_{i=1}^{N}(X_i-\overline{X})(X_i-\overline{X})^T = \widehat{\boldsymbol{\Psi}}, \tag{6.2}$$

where the mean vectors are computed from the N simulated sample of realizations from the posterior distribution. In general, when the strong law of large numbers is used to find numerical values of the expression on the left side of the equation (6.1) expressed in terms of a multiple integral, this technique is known as Monte Carlo integration.

Technical details on Monte Carlo integration methods may be found in Mode (2005) for the case of connecting HIV/AIDS models with data. In principle, the methods used in this paper could be extended such that these ideas could be applied to provide a method for estimating the parameters in the two-trait model under consideration. But this would be a major project which would entail not only the formulation a Bayesian model but also writing software to implement the algorithms designed to do the necessary computations.

Rather than following the Bayesian path, a decision was made to follow recent developments in sequencing genomes to make an assessment as to whether it would be possible to estimate parameters of a quantitative genetics model when regions of the genome could be used to develop working definitions of loci with at least two alleles at each locus in a set of loci. As can be seen by looking at the results in Chapters 3 and 4, it is straightforward to estimate genetic parameters by a direct approach for the case of either one trait or two or more traits. In particular, when several traits are under consideration, there is no need to subtract estimated covariance matrices, because, when working definitions of a set of loci with at least two alleles at each locus have been made, all covariance matrices may be estimated directly from the data, as is shown in Chapter 4.

APPENDIX

The purpose of this appendix is to prove that the formulas (3.15) and (3.16) are correct. As a first step in deriving a formula for the expectation of the sum of squares

$$Q_1 = b\sum_{i=1}^{a}(\overline{Y}_i - \overline{Y})^2, \tag{A.1}$$

it will be helpful to define some symbols that will lighten the notation. As will be shown, the formulas will expedite the derivation of $E[Q_1]$ and $E[Q_2]$.

To this end, let

$$Y_{i\cdot} = \sum_{j=1}^{b} Y_{ij}, \tag{A.2}$$

$$Y_{\cdot\cdot} = \sum_{i=1}^{a}\sum_{j=1}^{b} Y_{in}. \tag{A.3}$$

Then,

$$\overline{Y}_i = \frac{Y_{i\cdot}}{b}, \tag{A.4}$$

$$\overline{Y} = \frac{Y_{\cdot\cdot}}{ab}. \tag{A.5}$$

Therefore, in this notation,

$$\begin{aligned} Q_1 &= b\sum_{i=1}^{a}\left(\frac{Y_{i\cdot}}{b} - \frac{Y_{\cdot\cdot}}{ab}\right)^2 \\ &= \sum_{i=1}^{a}\frac{Y_{i\cdot}^2}{b} - 2b\sum_{i=1}^{a}\left(\frac{Y_{i.}}{b}\right)\left(\frac{Y_{\cdot\cdot}}{ab}\right) + ab\left(\frac{Y_{\cdot\cdot}}{ab}\right)^2 \\ &= \sum_{i=1}^{a}\frac{Y_{i\cdot}^2}{b} - 2\frac{Y_{\cdot\cdot}^2}{ab} + \frac{Y_{\cdot\cdot}^2}{ab} = \sum_{i=1}^{a}\frac{Y_{i\cdot}^2}{b} - \frac{Y_{\cdot\cdot}^2}{ab}. \end{aligned} \tag{A.6}$$

Similarly, it can be shown that

$$Q_2 = \sum_{i=1}^{a}\sum_{j=1}^{b}(Y_{ij} - \overline{Y}_i)^2 = \sum_{i=1}^{a}\sum_{j=1}^{b} Y_{ij}^2 - \frac{1}{b}\sum_{i=1}^{a} Y_{i\cdot}^2. \tag{A.7}$$

It can also be shown that

$$E\left[\overline{Y}_i^2\right] = \mu^2 + \alpha_i^2 + b\sigma_\epsilon^2. \tag{A.8}$$

Similarly, it can be shown that

$$\frac{E[Y_{\cdot\cdot}^2]}{ab} = ab\mu^2 + b\sigma_\alpha^2 + \sigma_\epsilon^2, \tag{A.9}$$

$$\frac{1}{b}\sum_{i=1}^{a} E\left[Y_{i\cdot}^2\right] = ab\mu^2 + ab\sigma_\alpha^2 + a\sigma_\epsilon^2. \tag{A.10}$$

Therefore,

$$
\begin{aligned}
E[Q_1] &= a\left(b\sigma_\alpha^2 + \sigma_\epsilon^2.\right) - \left(\sigma_\epsilon^2 + b\sigma_\alpha^2\right) \\
&= (a-1)\left(\sigma_\epsilon^2 + b\sigma_\alpha^2\right),
\end{aligned} \tag{A.11}
$$

which demonstrates that the formula (3.14) is valid.

Next, observe that

$$
\begin{aligned}
E&\left[\sum_{i=1}^{a}\sum_{j=1}^{b} Y_{ij}^2\right] \\
&= ab\left(\mu^2 + \sigma_\epsilon^2 + \sigma_\alpha^2\right).
\end{aligned} \tag{A.12}
$$

Hence,

$$
\begin{aligned}
E[Q_2] &= ab\left(\mu^2 + \sigma_\varepsilon^2 + \sigma_\alpha^2\right) - \left(ab\mu^2 + ab\sigma_\alpha^2 + ab\sigma_\epsilon^2\right) \\
&= ab\sigma_\varepsilon^2 - a\sigma_\varepsilon^2 \\
&= a(b-1)\sigma_\epsilon^2.
\end{aligned} \tag{A.13}
$$

Therefore, the formula (3.15) is valid.

REFERENCES

1. Anderson, T. W. (1958) An Introduction of Multivariate Statistical Analysis. John Wiley and Sons, New York.
2. Comstock, R. E. and Robinson, H. F. (1948) The components of genetic variation in populations of biparental progenies and their use in estimating the average degree of dominance. *Biometrics* **4**: 254–265.
3. Kempthorne, O. (1957) An Introduction to Genetics Statistics. John Wiley and Sons, New York, London.
4. Mode, C. J. and Robinson, H. F. (1959) Pleiotropism and the genetic variance and covariance. *Biometrics* **15**: 518–537.
5. Mode, C. J. (2005) A Bayesian Monte Carlo integration strategy for connecting stochastic models of HIV/AIDS with data. In: W. Y. Tan and H. Wu (eds.), Deterministic and Stochastic Models of AIDS Epidemics and HIV Infections with Interventions. World Scientific Chap. 3, pp. 61–76.

CHAPTER 2

On Fitting a Genetic Model to Data

BACKGROUND

While on a Philadelphia commuter train in the fall of 1970, I noted that the man who sat by me was reading a genetics journal. He introduced himself as Dr. David Gasser, Department of Medical Genetics in the School of Medicine of the University of Pennsylvania. It was mentioned that I was also a geneticist but, because of my interests in probability and statistics, as well as mathematics, I was a Professor of Mathematics at Drexel University. The campus of Drexel University shares a border with that of the University of Pennsylvania. Consequently, it was easy and convenient to meet and form a friendship as we talked about our mutual interests in genetics. It was mentioned that he had quantitative data on samples of blood from mice and thought a certain property of the blood was controlled by a single genetic factor. After thinking about the problem of testing whether a certain property of the blood of mice was indeed controlled by a single genetic factor, a nonparametric test was devised to test the hypothesis. Below is a paper that contains an account of a procedure for testing the hypothesis under consideration. The theme of this chapter is a departure from the theme of the book, because it does not deal with components of variance analysis. Nevertheless, even in this age of genomics, testing whether a genetic model fits the data is still of interest. A slightly revised version of the paper that resulted from this cooperative effort is shown below, which was published in *Mathematical Biosciences* in 1972; see Mode and Gasser (1972), *Mathematical Biosciences* **14**, 143–150.

A DISTRIBUTION-FREE TEST FOR MAJOR GENE DIFFERENCES IN QUANTITATIVE INHERITANCE

In this paper a new statistical test is proposed for detecting major gene differences in quantitative inheritance. It is based on the Smirnov distribution-free test and may be used to detect single gene differences between two inbred lines using data from parental, F., F2 and backcross generations. An application of the test is demonstrated using data from previously published experiments on the genetic control of specific antibody production.

1. Introduction

During the seven decades following the rediscovery of Mendel's laws in 1900, research in statistical methods for studying quantitative inheritance has developed along two lines. The first of those lines encompasses what may be called variance component methods, and from the historical point of view the famous paper of Fisher [1] has played an important role in the development of this line of research. There is an extensive and well-known literature on variance component analysis, so that only a few comments will suffice here. Briefly, in variance component analysis one partitions the total phenotypic variance into environmental and genetic components, and then by taking observations on sets of relatives such as full sibs or half sibs it is possible to obtain estimates of the environmental and genetic components of variance and thus estimate heritability. These methods of analysis seem to be most appropriate when the inheritance of a quantitative character is such that its variation in segregating populations cannot be explained in terms of genes at a few loci.

A second line of research in statistical methods for studying quantitative inheritance centers around fitting to data genetic models based on genes segregating at a few loci [2]. From the genetic point of view these methods have a great deal of intuitive appeal, for they attempt to explain quantitative variation in segregating populations in terms of the simpler models of qualitative genetics plus perhaps physiological considerations. Three examples of early papers in which attempts have been made to explain quantitative variation in segregating populations in terms of simple gene models are those of Nilsson-Ehle [3] and East [4] on quantitative characters in plants, and of Wright [5] on coat color in guinea pigs. More recently, Powers and Locke [6] and Powers [7] have attempted to explain quantitative variation in segregating populations by fitting to data certain genetic models involving three or four loci, using chi-square tests of goodness of fit. But unfortunately these tests of goodness of fit are not without difficulties. Some of the difficulties associated with these chi-square tests of goodness of fit as applied to problems in quantitative genetics are that they are based on assumptions of normality which are difficult to verify, the calculations are laborious, the question regarding the number of degrees of freedom associated with the test seems to be unsettled, and the test criterion is only approximately distributed as chi-square. In this paper it is shown that many of these difficulties may be circumvented by using some distribution-free test procedures, but before proceeding to the main parts of the paper at least one other difficulty of using simple genetic models to explain quantitative variations should be mentioned.

Perhaps the most serious objection to invoking simple genetic models to explain quantitative variation in segregating populations is that it is difficult to be sure that another model may not explain the data equally well. Suppose that we use some three-locus model to explain quantitative variation in a segregating population. Then how can one be sure that a four locus model or some other variation of a three-locus model might not explain the data equally well? It is beyond the scope of this paper to give any satisfactory answer to this question, but it is hoped that the simple statistical methods for detecting major gene differences discussed here will be helpful in unraveling more complex genetic situations.

The purpose of this paper is to present a nonparametric method of analysis as it applies to quantitative variation determined by single gene differences. Most quantitative characters are undoubtedly determined by multiple genetic loci, but by no means is it inconceivable that a single gene, perhaps in conjunction with environmental factors, is responsible for a metric trait. Even in some cases where several genes determine a quantitative character, it may be possible to isolate a single "leading factor" which is primarily responsible for the determination of that trait [2]. For example, quantitative differences are known to occur in the ability of some animals to mount a specific immune response, and in many cases these differences are determined by single genes [8]. In this paper a general method is presented which should allow an experimenter to determine whether a particular quantitative trait can or cannot be explained by the segregation of a single gene pair. As an illustration, the method has been applied to some previously published work in which it was shown that a specific immune response in certain cases is controlled by a single gene difference and in other cases by more than one gene.

2. A Brief Review of Theory

The distribution-free test which we shall utilize in this paper was proposed by Smirnov [9] and involves the concept of the empirical distribution function of a sample which we now describe. Suppose that a random variable X, representing the variation in some quantitative character, has the distribution function $F(x)$ so that

$$P[X \leq x] = F(x) \tag{1}$$

for all real x. Let $X_1, X_2, \ldots, X_n$ be a random sample of size n from a population with a distribution function given in [1] and let $Y_1 < Y_2, \ldots, Y_n$ be the sample values ordered from the smallest to the largest. Then $E(x)$, the empirical distribution function of the sample, is defined as follows. Put $E(x) = 0$ if $x < Y_1, E(x) = 1/n$ if $Y_1 \leq x < Y_2, E(x) = k/n$ if $Y_k \leq k/n < Y_{k+1}$ for $k = 2, \ldots, n-1$, and $E(X) = 1$ if $x \geq Y_N$. In other words, $E(x)$ merely determines the frequency of sample values less than or equal to the number x. Smirnov's procedure for testing whether two samples were drawn from the same population is as follows. Let X_1 be a random variable with distribution function $F_1[x]$, let X_2 be a random variable with distribution function $F_2(x)$ and let $E_1(x)$ and $E_2(x)$ be the empirical distribution functions of random samples of sizes m and n, respectively, drawn from populations with distribution functions $F_1(x)$ and $F_2(x)$. The criterion proposed by Smirnov for testing the hypothesis that $F_1(x) = F_2(x) = F(x)$ for all x for is the statistic

$$T = \sup_x |E_1(x) - E_2(x)|, \tag{2}$$

the largest vertical distance between the empirical distribution functions. The Smirnov statistic has the property that it is distribution-free under the hypothesis $F_1(x) = F_2(x) = F(x)$ for all x and $F(x)$ is continuous. If the reader is interested in a theoretical account of the Kolmogorov–Smirnov statistics, he may consult Birnbaum [10], Chapter 16. A very readable account of the applications of the Kolmogorov–Smirnov statistics to practical problems as well as tables giving their distributions has been given by Conover [10] (see Chapter 6 and pp. 397–401). It will be noted that the distribution of the statistic T has been tabulated for arbitrary values of m and n so that in applications we need not restrict ourselves to the case $m = n$.

We close this section by giving a brief numerical example illustrating the relative ease with which the Smirnov statistic may be calculated. For the sake of simplicity, suppose that a random sample of size two from the first population yields the values 1 and 3 and a random sample of size three from the second population yields the values 2, 4 and 6. Then $E_1(x) = 0$ if $x < 1$, $E_1(x) = 1/2$ and $E_1(x) = 1$ if $x > 3$. Similarly, $E_2(x) = 0$ if $x < 3$, $E_2(x) = 1/3$ if $2 \leq x < 4$, $E_2(x) = 2/3$ and $E_2(x) = 1$ if $x \geq 6$. The Smirnov test statistic for these two samples may be calculated by arranging the observations from both samples in their natural order, i.e. $1 < 2 << 3 < 4 < 5 < 6$, and then calculating $|E_1(x) - E_2(x)|$ on each of the intervals formed by the natural ordering of the observations. The calculations may be facilitated by using Table 2.1, where $D = |E_1(x) - E_2(x)|$. From this table it is easy to see that the Smirnov test statistic in this case is $T = 2/3$ and by consulting Table 17, p. 400 of Conover [11], the statistical significance of this value may be judged. For sample sizes $m = 2$ and $n = 3$ the critical value of T at the 20% level is 5/6. Therefore, since $T < 5/6$ in this case we would accept the hypothesis that the samples came from the same population at the 20% significance level.

Table 2.1. A Simple Numerical Example Illustrating the Calculation of the Smirnov Statistic

Intervals	$x < 1$	$1 \leq x < 2$	$2 \leq x < 3$	$3 \leq x < 4$	$4 \leq x < 6$	$x > 6$
$E_1(x)$	0	1/2	1/2	1	1	1
$E_2(x)$	0	0	1/3	1/3	2/3	1
D	0	1/2	1/6	2/3	1/3	0

3. An Application of the Smirnov Test

A frequently occurring situation in genetic research, particularly in small animal laboratories, is the case in which the experimenter takes observations with respect to some quantitative character on two parental inbred lines; an F_1 population formed by mating the inbred lines, and an F_2 population formed by mating individuals of the F_1 generation. The frequency distributions of the parental populations, the F_1 and the F_2, suggest that a single autosomal gene may differentiate the parents, but the variability in the F_1 population is such that it overlaps both parental populations, making it difficult to identify with certainty a putative F_1 genotype in the F_2 population. In such circumstances it would be very useful to have a simple statistical test to aid the experimenter in deciding whether a single autosomal gene did indeed differentiate the two inbred lines. Moreover, it would be very desirable if this test could be carried out with a minimum of calculations. The purpose of this section is to show that a simple experiment may be designed in such a way that the Smirnov statistic may be used to test the hypothesis that a single autosomal gene differentiates the parental lines, using only the assumption that the variability associated with each genotype may be characterized by some continuous but otherwise unknown distribution function.

Denote the parental lines by P_1 and P_2 and let the genotype of the P_1 population be AA and that of the P_2 population be aa so that the genotype of the F_1 generation is Aa. We assume that the variability in the P_1, P_2 and F_1 populations may be characterized in terms of the continuous distribution functions $F_1(x), F_2(x)$ and $F_3(x)$, respectively. Thus, if we let X_1, X_2 and X_3 be random variables describing the variation in the P_1, P_2 and F_1 populations, then

$$P\left[X_k \leq x\right] = F_k(x), \tag{3}$$

for $k = 1, 2, 3$. From the intuitive point of view the probabilities $P\left[X_k \leq x\right]$ for $k = 1, 2, 3$ may be interpreted as the frequencies of measurements less than or equal to x in the P_1, P_2 and F_1 populations. In the F_2 population the genotypes AA, aa and Aa occur with probabilities or frequencies $1/4, 1/4$ and $1/2$. Therefore, $F(x)$, the distribution function characterizing the variability in the F_2 population, is given by

$$F(x) = \frac{1}{4}F_1(x) + \frac{1}{4}F_2(x) + \frac{1}{2}F_3(x) \tag{4}$$

for all $x \in \mathbb{R}^+$, the set of positive measurements. In probability theory a distribution function of the form appearing (1) is known as a mixture, and taking observations on an F_2 population is equivalent to taking a random sample from a population whose distribution function is given by the equation (4). Since $F_1(x)$, $F_2(x)$ and F_3 are continuous by assumption, so is $F(x)$.

The Smirnov procedure can be applied to the problem of testing the hypothesis that a single autosomal gene differentiates the parental lines by essentially taking two random samples from a population whose distribution function is given by (4). The first of these samples could be a set of measurements on $m \geq 1$ F_2 generation individuals, and a second sample of size $n \geq 1$ could be obtained by sampling the P_1, P_2 and F_1 populations by a procedure which simulates an F_2 population under the hypothesis that a single autosomal locus with two alleles differentiates the two parental lines as indicated in [4]. For example, suppose that we toss two fair coins. Then, if we write H or T according as a coin falls, heads, or tails, the two coins may fall in four equally likely ways, namely HH, HT, TH or TT. If the coins fall HH or TT, then a random observation is taken from the P_1 or P_2 population, but if the coins fall either HT or TH, we take a random observation from the F_1 population. By repeating this process $n \geq 1$ times, we generate a random sample of size n from a population whose distribution function is given in equation [4]. Given these two random samples of size $m \geq 1$ and $n \geq 1$ the Smirnov statistic may be calculated by the procedure outlined in Section 2 to test the hypothesis that a single autosomal gene differentiates the inbred lines. Acceptance of the hypothesis that the two samples came from the same population could be used as evidence that a single autosomal gene did indeed differentiate the two inbred lines; while rejection of the hypothesis would suggest that a genetic model involving one locus was insufficient to explain the observed variation in the P_1, P_2, F_1 and F_2 populations.

An advantage of the procedure outlined above is that it allows an experimenter to use data already taken on P_1, P_2 and F_1 individuals, providing these individuals were grown under the same environmental conditions as the F_2 individuals. It will also be necessary to take into account the condition that each of these previously observed P_1, P_2 and F_1 populations contains a fixed number of individuals. For purposes of illustration, suppose that an experimenter has observed n_1, n_2 and n_3 individuals in the P_1, P_2 and F_1 populations and wishes to simulate an F_2 population by the sampling procedure outlined above. By fixing n in advance, it may happen that the observations in one of the populations are exhausted before a random sample of size n is obtained, and therefore the sampling procedure would not simulate a desired F_2 population, because at every step it is assumed that an observation is available from the P_1, P_2 or F_1 population. This difficulty can be easily overcome if the experimenter decides in advance to take $n = \min(n_1, n_2, n_3)$ observations. If sex differences are present, then of course each sex should be considered separately, and if the environmental conditions under which the P_1, P_2 and F_1 populations were grown differ from those of the F_2, then it would be necessary to design an experiment in which all individuals were grown under the same environmental conditions, with the sample sizes m and n being assigned in advance. The actual sample sizes, m and n, in small animal experiments will quite often be of the order from 10 to 20, but here, as is frequently the case elsewhere, the greater the number of observations, the greater the faith an experimenter can put in a statistical test.

A case of interest in which the Smirnov test could be very useful is in elucidating the genetics of the immune response. For example, it is well established that a single gene in mice determines the magnitude of the antibody response to a synthetic antigen called $(T, G) - A - L$. Many similar systems have been described, some unigenic and some multigenic, and there is a need for a simple statistical test which could aid the experimenter in deciding whether quantitative differences in the level of specific antibodies might be explained by the segregation of single genes.

When inbred strains of mice were immunized against the erythrocytic alloantigens determined by the $Ea - 1$ locus, there were highly significant differences in the amount of specific antibodies produced by the various strains. When the responding strain *YBR* was crossed with either of two non-responding strains, *BALE* and *CBA*, the data could be accounted for by the segregation of a single gene pair [12, 13]. In order to confirm the unigenic nature of this trait, animals from four successive backcross generations were tested, and in no case was there a significant deviation from the expected 1:1 ratio [13]. When *YBR* was crossed with the poorly responding strain $C57BL/10(BlO)$, the data could not be accounted for by single gene segregation [14], but appeared to be compatible with the segregation of two genetic loci [15]. We have used the data from these experiments to illustrate how the

Table 2.2. A Test to Determine Whether a Quantitative Trait (Response to Ea-I) Is Compatible with a Single Gene Hypothesis, Using Smirnov Statistics.

Generation	T	0.90	0.95	0.98
(a) $(YBR - CAB)F_1 \times YBR$	0.3900	0.3660	0.4080	0.4560
(b) $(YBR - CBA)F_1 \times CBA$	0.1667	0.4544	0.5066	0.5662
(c) $(YBR - CBA)F_2$	0.1551	0.3203	0.3571	0.3991
(d) $(YBR\_BALB)F_1 \times YBR$	0.1970	0.2942	0.3280	0.3666
(e) $(YBR - BALB)F_2$	0.1512	0.3137	0.349	0.3909
(f) $(YBR - B10)F_1 \times YBR$	0.2088	0.3157	0.3519	0.3933
(g) $(YBR - B10)F_1 \times B10$	0.3077	0.2581	0.2877	0.3216
(h) $(YBR - B10)F_2$	0.2374	0.1970	0.2160	0.2454

Smirnov test can be applied to test genetic hypotheses. The results of the test are shown in Table 2.2.

The intervals for the empirical distribution functions are based on the antibody titers of individual mice expressed as $-log_2[x]$ for $x \in \mathbb{R}$, the set of numbers recorded in each experiment. In each case $E_1(x)$ is the empirical distribution as observed experimentally, while $E_2(x)$ is the empirical distribution based on expectations from a single gene hypothesis and worked out according to the procedure described above; see the equation (4). The Smirnov statistic T is listed for each of the eight generations tested and the quantities w_p are shown for each of three confidence levels from a table of confidence levels for Smirnov tests. The null hypothesis is to be rejected at a given level of confidence if T exceeds w_p for that particular case. Interestingly enough, when $p = 0.95$, generations (a) through (e) are all compatible with single gene segregation, as shown by more traditional methods [12]. At the same level of confidence, the single gene hypothesis can be rejected for the generations shown in lines (g) and (h), which again corresponds to conclusions based on chi-square tests [14, 15].

By way of interpreting the first line of the table, the number $w_p = 0.3660$ in the third position is the confidence level of the test of the null hypothesis. In other words, in general, if p is the p value, then $w_p = 1 - p$ is the confidence level. Previous tests of genetic hypotheses on these data were based on arbitrary definitions of "responder" and "nonresponder" [12]. The test described in this paper circumvents the difficulty of making such arbitrary distinctions since no discrete classification of phenotypes is required.

4. A Look into the Future on Fitting Models to Data in an Era of Sequenced Genomes

It seems very likely that the genome of the mouse has been sequenced, and regions of the genome that have been found to be implicated in quantitative traits, such as the trait involving the blood of the mouse, have been considered in the foregoing sections of this chapter. Furthermore, it may be possible to characterize the notion of a major gene that has been discussed in the Mendelian paradigm in terms of regions of the genome that have been implicated to be involved in the expression of the trait. As times passes, it may also be possible to characterize the expression of a quantitative trait at the biochemical level involving coding regions or regions of a genome that influence the expression of a quantitative trait. Given such information it may be possible to formulate a statistical model that describes the expression of a quantitative trait and test whether it fits the data by using either the nonparametric methods described in this chapter and elsewhere or some parametric formulation. Even though it was mentioned specifically, a Monte Carlo simulation technique was used in testing the hypothesis under consideration; see Section 3 for details. If a more complex model were used, it would be possible to use a Monte Carlo simulation technique to test whether a proposed model is a good fit to the data.

REFERENCES

1. Fisher, R. A. (1918). *Trans. Roy. Soc. Edinb.* **52**: 399.
2. Wright, S. in Evolution and the Genetics of Populations, (Vol. 1), University of Chicago, Chicago (1968), pp. 373–420.
3. Nilsson-Ehle, H. (1908). *Botaniska Notiser* **1908**: 257.
4. East, E. M. (1916). *Genetics* **1**: 164.
5. Wright, S. (1917). *J. Hered.* **8**: 244.
6. Powers, L. and Locke, L. F. (1950). *USDA Tech. Bull. No.* **998**.
7. Powers, L. (1955). *USDA Tech. Bull. No.* **1131**.
8. McDevitt, H. O. and Benacerraf, B. (1969). *Adv. Immunol.* **11**: 31.
9. Smirnov, N. W. (1939). *Bull. Moscow Univ.* **2**: 3.
10. Birnbaum, Z. W. (1962). *Introduction to Probability and Mathematical Statistics*, Harper, New York.
11. Conover, W. J. (1971). *Practical Nonparametric Statistics*, Wiley, New York.

12. Gasser, D. L. (1969). *J. lmmunol.* **103**: 66.
13. Gasser, D. L. (1970). *Ph.D. thesis*, The University of Michigan.
14. Gasser, D. L. (1970). *J. Immunol.* **105**: 908.
15. Gasser, D. L. and Shreffler, D. C. (1972). *Nat. New Biol.* **235**: 155.

CHAPTER 3

Estimating Effects and Variance Components in Models of Quantitative Genetics in an Era of Sequenced Genomes*

ABSTRACT

As in many other areas of research in genetics, the availability of sequenced genomes in samples of individuals has revolutionized the study of quantitative traits, because researches have developed statistical evidence regarding the locations of genomic regions, loci, that have been implicated with the expression of a quantitative trait or traits. Therefore, in cases in which it is possible to develop operational definitions of at least two alleles at each locus, genomic regions, it becomes possible to identify the genotype of each individual with respect to a set of loci that have been shown in other experiments to influence the expression of a quantitative trait. As will be shown in this paper, by knowing the genotype of each individual in a sample with respect to a set of identified loci, it is now possible to directly estimate effects that are measures of not only intra-allelic interactions at each locus under consideration but also various types of epistatic effects that are measures of interactions among alleles at different loci, governing the inheritance of a quantitative trait. These straightforward methods of estimation differ from those used in classical quantitative genetics, because such effects and corresponding variance components could be estimated indirectly, using

*This chapter was published as a paper in an online journal: Mode, C. J. (2014) Estimating effects and variance components in models of quantitative genetics in an era of sequenced genomes. *Global Journal of Science Frontier Research — Mathematics and Decision Sciences.* Issue 5, Version 1.0, Online ISSN 2249-4626. The content of this chapter is a PDF image of the LaTeX version of the paper that was accepted for publication.

analysis of variance procedures or some version of general linear models that have been and are widely used in statistical genetics. The direct method of estimation described in this paper shows promise towards shifting the working paradigm that has been used in classical models of the genetics of quantitative traits involving the estimation of variance components to a more direct approach and simpler approach.

1. Introduction

In an interesting review paper by Stranger *et al.* (2010) [21], the impact of genome-wide association studies on the genetics of complex traits is discussed in depth. Among these complex traits are Alzheimer's disease (AD) and immune-mediated diseases such as rheumatoid arthritis. For the case of AD, in a recent paper Raj *et al.* (2012) [19] have reported that 11 regions of the human genome are involved in susceptibility to this disease, and, moreover, there is evidence that four of these regions form a protein network that is under natural selection. Similarly, in paper by Rossin *et al.* (2011) [20], it has been found that proteins encoded in genomic regions associated with immune-mediated disease physically interact and this interaction may also suggest some basic biological mechanisms underlying such diseases.

There is also another technological development, the sequencing of entire genomes of individuals, that may lead to a deeper understanding of the relationships of phenotypes to genotypes. Suppose, for example, that a sample of individuals with symptoms of a disease, such as AD, is available and that the genome of each individual in the sample has been sequenced. Furthermore, suppose that some quantitative measurement is made on each individual in the sample. These measurements will vary among individuals and let W denote a random variable characterizing this variation. Given that the genome of each individual in the sample has been sequenced, the genotype of each individual in the sample can, in principle, be identified with respect to the 11 loci under consideration for the case of AD. It will also be supposed that at each locus at least two alleles can be identified.

In classical quantitative genetics, the loci and alleles at each locus were treated abstractly, because an investigator did not, in general, know the location of the hypothesized loci in the genome of a species or the number of alleles at each locus. However, for the case of AD cited above, the genotype of each individual in the sample can be identified with respect to each of the 11 loci, and in some cases it may be known with respect to combinations of the 11 loci or even all 11 loci. Such technological developments provide opportunities to extend some of the ideas of classical quantitative genetics into the age of sequenced genomes. Moreover, as will be demonstrated in subsequent sections of this paper, when the genotype of each individual in a sample is known, the estimation of parameters of the model may be carried out in a relatively simple and straightforward manner based on elementary methods of statistical estimation.

As is recognized among many who have worked in the field of quantitative genetics, the subject known as components of variance analysis began with the publication of a paper on correlations among relatives on the supposition of Mendelian inheritance by R. A. Fisher (1918) [7]. In his paper, Fisher attempted to reconcile existing biometrical theories with Mendelian genetics, which led him to describe genetic variation in terms of components of variance. During the 1950s, other investigators published papers that were motivated by the paper by Fisher. Among these investigators was Kempthorne (1954) [10], who introduced an approach to components of variance analysis based on effects defined in terms of expectations of genetics values with respect to the genotypic distribution under the assumption that the population was in a Hardy–Weinberg equilibrium. An alternative approach was introduced by Cockerham (1954) [5] and it is also of historical interest, because it contains an extensions of Fisher's ideas to accommodate epistatic effects in terms of ideas depending on the concept of orthogonality. If the reader is interested in further details and development of the ideas of Fisher and other workers, it is suggested that the book Kempthorne (1957) [11] be consulted, where many of the themes of statistical genetics as they existed during the 1950s were summarized and extended.

The techniques introduced in these papers have also been applied in the current genomic era. Examples of the ideas introduced by Cockerham have been applied in the paper Kao *et al.* (2002) [9], and those of Kempthorne have been applied and extended in the paper Mao *et al.* (2006) [15]. The ideas of Kempthorne were also used and extended in the paper of Mode and Robinson (1959) [16] as well as in unpublished lecture notes by the author written and presented during the period 1960 to 1966. Furthermore, the roots of the ideas presented in this paper are extensions of some of the unpublished material in the lecture notes complied by the author during the period 1960 to 1966.

During the years following Fisher's seminal work, an extensive literature on quantitative genetics has evolved. It is beyond the scope of this paper to review this literature and in what follows a few books on the subject will be cited. A book that has been very popular with quantitative geneticists is that of Falconer and MacKay (1996) [6] as well as earlier editions. Another book of interest on quantitative genetics is that of Bulmer (1980) [4]. Both of these books contain extensive lists of references on quantitative genetics. A more recent book on genetics and analysis of quantitative traits is that of Lynch and Walsh (1998) [14]. This influential tome consists of over 900 pages and contains what seems to be the most extensive treatment of the subject of quantitative genetics published in the 20th century. The principal focus of this book is a biological and evolutionary point of view along with an extensive use of applied statistical methods. There is also an extensive list of papers on quantitative genetics that a reader who is interested in quantitative genetics may wish to peruse. The book by Liu [13] on statistical genetics focuses on statistical genetics along with linkage, mapping and quantitative trait linkage (QTL) analysis. Two recent books on statistical genetics are those of Laird and Lange [7] and Wu, Ma and Casella (2010) [22].

Historically, procedures for estimating components of the genetic and environmental variances have been based on experimental designs or observational data involving various types of relatives. If the reader is interested in an account of such experimental designs, it is suggested that Chapter 6 of Bulmer (1980) [4] be consulted. An in-depth account of estimation procedures in various genetic settings may be found in Section III of the book by Lynch and Walsh (1998) [14]. In this paper, however, it will be shown that when the genotype of each individual in a sample is known at the DNA level, it is possible to estimate various types of genetic parameters directly, including variance components, using elementary statistical ideas. It should also be mentioned that the ideas presented in this paper are extensions of techniques from unpublished notes on quantitative genetics written by the author during the period 1960 to 1966. In these notes, it was assumed that for the one-locus case the population was in a Hardy–Weinberg equilibrium, and for the case of multiple loci, it was assumed that the population was in linkage equilibrium. In this paper, however, these assumptions have been relaxed.

When two or more quantitative traits are under consideration several measurements are taken on each individual. In this case, it is assumed that the autosomal loci under consideration may influence the expression of alleles for two or more traits. In classical genetics, such joint expressions of alleles for quantitative or qualitative traits are referred to as pleiotropism. In a recent paper, Mode (2014) [17], this case has been worked out in detail.

2. The Case of One Locus with Multiple Alleles

Let $\mathbb{A}$ denote a finite set of alleles at some autosomal locus in a diploid species such as man. Elements of $\mathbb{A}$ will be denoted by the symbols x and y, and the genotype of an individual with respect to the locus will be denoted by (x, y), where $x \in \mathbb{A}$ and $y \in \mathbb{A}$ denote, respectively, the alleles contributed by the maternal and the paternal parent of the individual under consideration. As the technology underlying the sequencing of DNA evolves, it seems likely that it will be possible to distinguish the DNA contributed by each parent to an offspring. More precisely, let $\mathbb{A} \times \mathbb{A}$ denote the Cartesian product of the set $\mathbb{A}$ with itself. Then $\mathbb{G} = \mathbb{A} \times \mathbb{A}$ is the set of all possible genotypes at the locus under consideration and $(x, y) \in \mathbb{G}$ for every genotype (x, y).

One of the objectives in formulating models in quantitative genetics is to provide a framework such that phenotypic measurements on a population of individuals may formally be connected with the genotype of each individual. For many decades it has been observed that phenotypic measurements among individuals with the same genotype in a given environment may vary. But it has also been observed that in populations consisting of several genotypes responses of the genotypes to a given environment may also vary. Let W denote a random variable that takes values in the set $\mathbb{R}_W$ of real numbers that constitute the set of possible phenotypic measurements of individuals in the population. In general, by assumption, the numbers in the set $\mathbb{R}_W$ will depend on the genotype of a homogeneous set of individuals.

Given a genotype $(x,y) \in \mathbb{G}$, let $f(w \mid (x,y))$ denote the conditional probability density function of the random variable W. Then,

$$E[W \mid (x,y)] = \mu(x,y) = \int_{\mathbb{R}_W} wf(w \mid (x,y))dw \tag{2.1}$$

is the conditional expectation of the random variable W, given the genotype (x,y). It will be assumed that $\mu(x,y)$ is finite for all genotypes $(x,y) \in \mathbb{G}$. Let $p(x,y)$ denote the probability frequency, that an individual chosen at random from the population is of genotype (x,y). Then, the unconditional expectation of the random variable W is, by definition,

$$\mu = E[W] = \sum_{(x,y)} p(x,y)E[W|(x,y)] = \sum_{(x,y)} p(x,y)\mu(x,y). \tag{2.2}$$

It is assumed that $p(x,y) \geq 0$ for all $(x,y) \in \mathbb{G}$ and

$$\sum_{(x,y)} p(x,y) = 1. \tag{2.3}$$

In what follows, it will also be helpful to observe that the joint distribution of a random genotype (x,y) and the phenotypic random variable W is $g((x,y),w) = p(x,y)f(w|(x,y))$ for all $(x,y) \in \mathbb{G}$ and $w \in \mathbb{R}_W$.

Next, observe that the equation

$$W = \mu + \big(\mu(x,y) - \mu\big) + \big(W - \mu(x,y)\big) \tag{2.4}$$

is valid and provides a linear relationship connecting an observed phenotypic measure W with the expectation μ, a measurement of a genetic effect expressed by the deviation $(\mu(x,y) - \mu)$ and the term $(Z - \mu(x,y))$, which may be interpreted as a measure of deviation of the phenotypic measure W from $\mu(x,y)$ due to environmental conditions. By definition, the total phenotypic variance in the population is

$$\mathrm{var}_P[W] = E\big[(W - \mu)^2\big]. \tag{2.5}$$

A widely used technique in quantitative genetics is to partition the total phenotypic variance into a genotypic variance measuring the variation among genotypes in their responses to environmental conditions and an environmental variance measuring the variation of the phenotypic measure W around the genotypic values $\mu(x,y)$ for every $(x,y) \in \mathbb{G}$. From equation (2.4), it follows that

$$(W - \mu)^2 = \left(\mu(x,y) - \mu\right)^2 + \left(W - \mu(x,y)\right)^2 + 2\left(\mu(x,y) - \mu\right)\left(W - \mu(x,y)\right). \tag{2.6}$$

Therefore,

$$E\left[(W - \mu)^2 \,|(x,y)\right] = \left(\mu(x,y) - \mu\right)^2 + \left(W - \mu(x,y)\right)^2 + 2\left(\mu(x,y) - \mu\right) E\left[\left(W - \mu(x,y)\right) |(x,y)\right]. \tag{2.7}$$

But

$$E\left[\left(W - \mu(x,y)\right) |(x,y)\right] = 0. \tag{2.8}$$

Therefore,

$$E\left[(W - \mu)^2 \,|(x,y)\right] = \left(\mu(x,y) - \mu\right)^2 + \left(W - \mu(x,y)\right)^2. \tag{2.9}$$

In deriving equation (2.9), some well-known properties of conditional expectations have been used. Namely, for any function $f(x)$ with the domain a subset of $\mathbb{R}$, the set of real numbers, and the range a subset of $\mathbb{R}$, it follows from well-known properties of conditional expectations that $E\left[f(X)|X\right] = f(X)$. It is also well known that if two random variables X and Y with range $\mathbb{R}$ are under consideration, then $E\left[XY|X\right] = XE[Y|X]$.

An equivalent representation of the phenotypic variance in equation (2.5) is

$$\text{var}_P[W] = \sum_{(x,y)} p(x,y) E\left[(W - \mu)^2 \,|(x,y)\right]. \tag{2.10}$$

Given (2.9), it seems reasonable to define the genetic variance due to genetic effects in the population as

$$\text{var}_G[W] = \sum_{(x,y)} p(x,y) \left(\mu(x,y) - \mu\right)^2. \tag{2.11}$$

Similarly, the variance due to environmental effects is, by definition,

$$\text{var}_E[W] = \sum_{(x,y)} p(x,y) E[(W - \mu(x,y))^2 |(x,y)]. \tag{2.12}$$

From equation (2.9) it follows, therefore, that

$$\mathrm{var}_P[W] = \mathrm{var}_G[W] + \mathrm{var}_E[W]. \tag{2.13}$$

At this point in the development of the content of this paper, it should be mentioned that equation (2.13) is not new to quantitative genetics, but its derivation is a departure from derivations that appeared in some papers and books on the subject. For example, in some formulations the effects $(\mu(x,y) - \mu)$ and $(W - \mu(x,y))$ are treated as abstract uncorrelated random variables and sometimes it is assumed that genetic and environmental effects are independent. But it follows from the use of conditional expectations that these types of assumptions are not necessary in the derivation of (2.13).

From equation (2.13), it can be seen that the total phenotypic variance may be partitioned into two component variances, namely the genetic and environmental variance. If both sides of equation (2.13) are divided by the phenotypic variance, then it is easy to see that

$$1 = F_G + F_E,$$

where

$$F_G = \frac{\mathrm{var}_G[Z]}{\mathrm{var}_P[Z]}$$

and

$$F_E = \frac{\mathrm{var}_E[Z]}{\mathrm{var}_P[Z]}. \tag{2.14}$$

Some authors refer to F_G as a measure of the heritability of the quantitative trait. In what follows, F_G will be denoted by H_G and referred to as a measure of heritability.

This latter ratio has been given various names by authors of books on quantitative genetics. For example, in the book by Falconer and Mackay (1996) [6], on page 123, an expression similar to the ratio H_G is called the degree of genetic determination. Other authors, such as Wu *et al.* (2010) [22], refer to this ratio as heritability in the broad sense and provide an example in which this parameter may be estimated by an analysis of variance procedure based on a designed breeding experiment; see page 178. If the reader is interested in pursuing the subject of heritability further, it is suggested that the book by Liu (1998) [13] be consulted; in particular, see pages 34 and 35. A more in-depth treatment of the concept of heritability may be found in the book by Lynch and Walsh (1998) [14]; see pages 170–175 and elsewhere in that book.

Several recent papers have also been devoted to applications of the concept of heritability. Among these papers is that of Zaitlen *et al.* (2013) [24], who use extended genealogies to estimate components of heritability for 23 quantitative and dichotomous traits, using closely and distantly related relatives. In an interesting paper Price *et al.* (2011 [11] estimated variance components using single tissue data on cross-tissue heritability gene expression on individuals related by descent and also unrelated individuals. In a paper by Yang *et al.* (2010) [23] the heritability of human height is studied. These authors show that by considering all common SNPs simultaneously, 45% of the phenotypic variance in human height can be attributed to genetic variation. It should be mentioned, however, that the quantitative model or models used by these authors were not as comprehensive as the structure that will be developed in subsequent sections of this paper.

In a subsequent section of this paper, procedures for estimating the components of variance just defined will be presented. It is recognized, however, that an investigator may be interested in testing statistical hypotheses as to whether the expectations and variances among the genotypes do indeed differ, but a discussion of tests of hypotheses is beyond the scope of this paper, which will be limited to a presentation of straightforward procedures for estimating the components of variance defined above.

Before proceeding to a discussion of estimation procedures, however, it is interesting to note that, even though the rhetoric in this section was confined to the case of one autosomal locus with multiple alleles, the formulas can be easily extended to the case of some finite number of autosomal loci $n \geq 2$ with a finite number of alleles at each locus. For let (x, y) denote the genotype of an individual with respect to n loci, where $x = (x_1, x_2, \ldots, x_n)$ and $y = (y_1, y_2, \ldots, y_n)$ denote, respectively, the alleles inherited from the maternal and the paternal parent, and suppose that $p(x, y)$ is the probability of selecting an individual of genotype (x, y) at random in the population. Then, it is easy to show that equation (2.13) also holds for some number $n \geq 2$ of autosomal loci, but the details of proving this statement will be left as an exercise for the reader.

Given that it can be shown that the equation (2.13) holds for any number of loci $n \geq 2$, it is interesting to note that with respect to AD the total phenotypic variance may be partitioned into the genetic and environmental components for any combination of the 11 loci that have been implicated with this disease. In particular, it would be of interest to estimate the heritability for each of the 11 loci or in combinations of loci in order to gain some insights as to whether heritability would increase as the number of loci under consideration increases.

3. A Partition of the Genetic Variance into the Additive and Intra-allelelic Components for the Case of One Autosomal Locus

Again let $p(x, y)$ denote the probability of finding an individual of genotype (x, y) a population. This probability is also known as the frequency of genotype (x, y) in a population. In most past formulations of models in quantitative genetics for the one-locus case, it has been assumed that a population was in a Hardy–Weinberg equilibrium and there was no mutation or selection. Mutation and selection will not be considered in this paper, but the condition that a population is in a Hardy–Weinberg equilibrium will be relaxed. Let $p(x)$ and $p(y)$ denote, respectively, the frequencies of alleles x and y in a population. Then, a population is in a Hardy–Weinberg equilibrium if $p(x, y) = p(x)p(y)$ for all genotypes $(x, y) \in \mathbb{G}$. In this section, the condition that a population is in a Hardy–Weinberg equilibrium will not be assumed, because in many populations this assumption may not hold. It should be mentioned, however, that an investigator may wish to test whether a sample from a population passes a statistical test or tests for a Hardy–Weinberg equilibrium.

To relax the assumption that a population is in a Hardy–Weinberg equilibrium, it will be necessary to deal with conditional probabilities and expectations. Let

$$p(x) = \sum_{y} p(x, y) \tag{3.1}$$

denote that marginal distribution for all maternal alleles $x \in \mathbb{A}$ in a population, and similarly let

$$p(y) = \sum_{x} p(x, y) \tag{3.2}$$

denote the marginal distribution for all paternal alleles $y \in \mathbb{A}$. Then, if $p(y) \neq 0$,

$$p(x|y) = \frac{p(x,y)}{p(y)} \tag{3.3}$$

is the conditional distribution of the alleles $x \in \mathbb{A}$, given allele $y \in \mathbb{A}$. Similarly, if $p(x) \neq 0$, then

$$p(y|x) = \frac{p(x,y)}{p(x)} \tag{3.4}$$

is the conditional distribution of alleles $y \in \mathbb{A}$, given allele $x \in \mathbb{A}$. The formulas just derived may be summarized in the equation

$$p(x,y) = p(x)p(y|x) = p(y)p(x|y) \tag{3.5}$$

for all genotypes $(x,y) \in \mathbb{G}$.

Therefore, the conditional expectation of $\mu(x,y)$, given x, is by definition

$$\mu(x) = \sum_{y} p(y|x)\mu(x,y). \tag{3.6}$$

Thus, the unconditional expectation of $\mu(x)$ is

$$E[\mu(x)] = \sum_{x} p(x)\mu(x) = \sum_{x}\sum_{y} p(x,y)\mu(x,y) = \mu. \tag{3.7}$$

For a justification of this equation, see equation (2.5). In particular, if the population is in a Hardy–Weinberg equilibrium, then $p(x,y) = p(x)p(y)$ for all $(x,y) \in \mathbb{G}$ and the equation (3.6) becomes

$$\mu(x) = \sum_{y} p(y|x)\mu(x,y) = \sum_{y} p(y)\mu(x,y), \tag{3.8}$$

because in this case $p(y|x) = p(y)$. Thus, in formulations in which the assumption that a population is in a Hardy–Weinberg equilibrium is in force, (3.8) is the definition of the average value of maternal allele x in a population. Similarly, by using techniques similar to those used in the derivation of a formula for $\mu(x)$, it is straightforward to derive a formula for $\mu(y)$, the average value for paternal allele y in the population that is not in a Hardy–Weinberg equilibrium.

To cast the formulation in terms of an analysis of variance structure, it is useful to define the effects of alleles x and y as the deviations

$$\alpha(x) = \mu(x) - \mu,$$
$$\alpha(y) = \mu(y) - \mu. \tag{3.9}$$

Observe that the unconditional expectations of these deviations is $E[\alpha(x)] = E[\alpha(y)] = 0$. The deviation

$$\alpha(x, y) = \mu(x, y) - \mu - \alpha(x) - \alpha(y) \tag{3.10}$$

is a measure of interactions among the maternal and paternal alleles. In this case, it is also easy to see that the unconditional expectation of this deviation is $E[\alpha(x, y)] = 0$. Alternatively, the deviations just described can be written in the form of an analysis of the variance equation

$$\mu(x, y) = \mu + \alpha(x) + \alpha(y) + \alpha(x, y), \tag{3.11}$$

which holds for all genotypes $(x, y) \in \mathbb{G}$. This equation suggests that it seems reasonable to call the terms $\alpha(x)$ and $\alpha(y)$ the additive effects of alleles. With the exception of μ, the terms on the right side of equation (3.11) are known statistically as effects. For if $\alpha(x, y) = 0$ for all genotypes, then

$$\mu(x, y) = \mu + \alpha(x) + \alpha(y) \tag{3.12}$$

for all $(x, y) \in \mathbb{G}$ so that the effects of alleles x and y have an additive effect on the expectation $\mu(x, y)$. But, if $\alpha(x, y) \neq 0$ for all (x, y), then there are interactions among the maternal and paternal alleles.

Having defined additive and intra-allelic interaction effects, the next step in the formulation is to define the additive and intra-allelic interaction variances. The additive genetic variance in the population is defined by

$$\text{var}_G(A) = \sum_x p(x)\alpha^2(x) + \sum_y p(y)\alpha^2(y), \tag{3.13}$$

and the intra-allelic interaction, IAI, variance is defined by

$$\text{var}_G(\text{IAI}) = \sum_{(x,y)} p(x, y)\alpha^2(x, y). \tag{3.14}$$

To connect these variances with the total genetic variance in a population, write equation (3.11) in the form

$$\mu(x, y) - \mu = \alpha(x) + \alpha(y) + \alpha(x, y)$$

and square both sides. The result is

$$(\mu(x,y)-\mu)^2=\alpha^2(x)+\alpha^2(y)+\alpha^2(x,y)+R(x,y), \tag{3.15}$$

where

$$R(x,y)=2\alpha(x)\alpha(y)+2\alpha(x)\alpha(x,y)+2\alpha(y)\alpha(x,y). \tag{3.16}$$

By multiplying equation (3.15) by $p(x,y)$ and summing over all genotypes (x,y), it follows that

$$\mathrm{var}_G[W]=\mathrm{var}_G(A)+\mathrm{var}_G(\mathrm{IAI})+E[R(x,y)], \tag{3.17}$$

where

$$E[R(x,y)]=T_1+T_2+T_3. \tag{3.18}$$

The explicit forms of the symbols on the right, which involve covariances, are as follows:

$$T_1=2\sum_{(x,y)}p(x,y)\alpha(x)\alpha(y),$$

$$T_2=2\sum_{(x,y)}p(x,y)\alpha(x)\alpha(x,y),$$

$$T_3=2\sum_{(x,y)}p(x,y)\alpha(y)\alpha(x,y). \tag{3.19}$$

In general, $E[R(x,y)]\neq 0$, but there is a case where $E[R(x,y)]=0$. Suppose that the population is in a Hardy–Weinberg equilibrium so the $p(x,y)=p(x)p(y)$ for all genotypes $(x,y)\in\mathbb{G}$. Then, T_1 may be written in the form

$$T_1=2\left(\sum_x p(x)\alpha(x)\right)\left(\sum_y p(y)\alpha(y)\right). \tag{3.20}$$

But

$$\sum_x p(x)\alpha(x)=0,$$

and therefore $T_1=0$. Similarly, T_2 may be written in the form

$$T_2=2\left(\sum_x p(x)\alpha(x)\right)\left(\sum_y p(y)\alpha(x,y)\right). \tag{3.21}$$

For every fixed x, consider

$$\sum_y p(y)\alpha(x,y) = \sum_y p(y)\left(\mu(x,y) - \mu - \alpha(x) - \alpha(y)\right)$$
$$= \alpha(x) - \alpha(x) - \sum_y p(y)\alpha(y) = 0, \quad (3.22)$$

for every $x \in \mathbb{A}$. Therefore, $T_2 = 0$, and by a similar argument it can be shown that $T_3 = 0$ so that $E\left[R(x,y)\right] = 0$.

Thus, for the case a population is in a Hardy–Weinberg equilibrium at some autosomal locus, it follows that the total genetic variance may be partitioned into the additive and intra-allelic interaction variances. In symbols,

$$\mathrm{var}_G[W] = \mathrm{var}_G(A) + \mathrm{var}_G(\mathrm{IAI}). \quad (3.23)$$

It is interesting to observe that when the genotype of each individual may be identified, each of the component variances on the right may be estimated separately. But, before the age of genomics, in quantitative genetic studies, the genotype of each individual in a population could not be identified. Under such circumstances, experiments could be designed in such a way that components of variance in equation (3.23) could be estimated from mean squares in an analysis of variance table. It should be noted, however, when the effects $\alpha(x), \alpha(y)$ and $\alpha(x,y)$ can be estimated from the data, all the covariances terms in $E[R(x,y)]$ could also be estimated. In such cases, one could also estimate the term $E[R(x,y)]$ in equation (3.17), which would be of interest in its own right for the cases in which the population was not in a Hardy–Weinberg equilibrium at the autosomal locus under consideration.

There is a notationally more succinct way to represent the variances and covariances encountered in the above discussion. For each genotype $(x,y) \in \mathbb{G}$ let

$$\mathbf{\Phi}(x,y) = \begin{pmatrix} \alpha(x) \\ \alpha(y) \\ \alpha(x,y) \end{pmatrix} \quad (3.24)$$

denote a 3×1 matrix whose elements are defined above. The transpose of this matrix is

$$\mathbf{\Phi}^T(x,y) = \left(\alpha(x) \quad \alpha(y) \quad \alpha(x,y)\right). \quad (3.25)$$

Next, observe that

$$\boldsymbol{\Psi}(x, y) = \boldsymbol{\Phi}(x, y)\boldsymbol{\Phi}^T(x, y) \tag{3.26}$$

is a 3×3 matrix and the element in position $(1, 1)$ is $\alpha^2(x)$, the element in position $(1, 2)$ is $\alpha(x)\alpha(y)$ and, by proceeding in this way, all nine of the elements in the matrix $\boldsymbol{\Psi}(x, y)$ as squares or products of the elements in the vector $\boldsymbol{\Phi}(x, y)$. Let $\boldsymbol{\Psi}_G$ denote the 3×3 genetic variance–covarince matrix for the autosomal locus under consideration. Then,

$$\boldsymbol{\Psi}_G = \sum_{(x,y)} p(x, y)\boldsymbol{\Psi}(x, y). \tag{3.27}$$

From now on $\boldsymbol{\Psi}_G$ will be called the genetic covariance matrix for the autosomal locus under consideration. It should be observed that the variance components on the right of equation (3.13) are in the principal diagonal positions $(1, 1)$ and $(2, 2)$ of the matrix $\boldsymbol{\Psi}_G$. Moreover, the sum of all elements of the principal diagonal of this matrix is the term $E[R(x, y)]$ in (3.17). Given the genetic matrix $\boldsymbol{\Psi}_G$, it may be useful to compute the eigenvalues of this matrix as well as its principal components in addition to estimating the components of the matrix $\boldsymbol{\Psi}_G$. As will be seen in subsequent sections, the matrix approach to computing the genetic covariance matrix described in this section will make it possible to describe the computation of the genetic covariance matrix for cases in which more than one autosomal locus is under consideration.

It can be seen from a perusal of books on statistical genetics that the approach used in this section and in subsequent sections of this paper to partition the genetic variance into components differs from that used in some books cited in the introduction. For example, on page 54 of the book Laird and Lange (2011) [12] a phenotypic measurement Y of a quantitative trait is represented as a linear combination of unknown parameters with indicator functions coefficients plus a random error term. Included in these terms are parameters for the additive effects of alleles as well as a codominant effect, which appears to be related to the intra-allelic interaction term defined this section. Such models appear to belong to the class of generalized linear models that are widely used in numerous areas of applied statistics. In particular, in the books on statistical genetics cited in the introduction, linear models similar to that cited in Laird and Lange (2011) have been used. As can be seen from the derivations presented in this section, however, the additive and interaction effects of alleles in the case of one autosomal locus are defined in terms of conditional expectations with respect to the genotypic distribution. Moreover, this scheme of using conditional expectations in defining effects when partitioning the total genetic variance into components will be used extensively in subsequent sections of this paper and provides a methodology for estimating effects and corresponding variance components directly from data. Furthermore, as will be shown subsequently, the squared effects making up a component of variance may also be estimated directly from a data set such that the genomes of all individuals in this sample have been sequenced.

4. Estimation of Parameters and Effects from Data

In this section, a procedure for estimating the parameters defined in the forgoing sections from phenotypic data will be outlined. Suppose that in a sample of individuals, $n((x,y)) \geq 2$ individuals of genotype $(x,y) \in \mathbb{G}$ are observed and let the random variables $W_\nu(x,y)$, for $\nu = 1, 2, \ldots, n(x,y)$, denote a sample of phenotypic measurements on the $n((x,y))$ individuals of genotype (x,y) with respect to some quantitative trait. Usually, the set of phenotypic measurements will belong to some set $\mathbb{R}$ of continuous real numbers; it will also be supposed that these random variables are independently and identically distributed according to a common but unknown distribution with a finite expectation and variance. Let

$$n = \sum_{(x,y)} n(x,y) \tag{4.1}$$

denote the total number of individuals in the sample, where the sum runs over all genotypes $(x,y) \in \mathbb{G}$. Then, the random variable

$$\widehat{p}(n(x,y)) = \frac{n(x,y)}{n} \tag{4.2}$$

is an estimator of the frequency $p(x,y)$ of genotype (x,y) in a population or subpopulation from which a sample of individuals was drawn. It is interesting to note that if it is assumed that the numbers $n(x,y)$ for $(x,y) \in \mathbb{G}$ are viewed as realizations from a multinomial distribution with probabilities $p(x,y)$ for

$(x, y) \in \mathbb{G}$ and sample size n, then $E[\widehat{p}(n(x, y))] = np(x, y)/n = p(x, y)$ so that $\widehat{p}(n(x, y))$ is an unbiased estimator of $p(x, y)$ for all $(x, y) \in \mathbb{G}$.

Similarly, the random variable

$$\widehat{\mu}(x, y) = \frac{1}{n(x, y)} \sum_{\nu=1}^{n(x,y)} W_{\nu}(i, j) \tag{4.3}$$

is an estimator of the parameter $\mu(x, y)$. This estimator is conditionally unbiased, because $\widehat{E}[\widehat{\mu}(x, y)|(x, y)] = n(x, y)\mu(x, y)/n(x, y) = \mu(x, y)$ for all genotypes $(x, y) \in \mathbb{G}$. Therefore, the random variable

$$\widehat{\mu} = \sum_{(x,y)} \widehat{p}(n(x, y))\widehat{\mu}(x, y) \tag{4.4}$$

is an estimator of the parameter μ. From these definitions, it follows that the random variable

$$\widehat{\text{var}}_G[W] = \sum_{(x,y)} \widehat{p}(x, y) \left(\widehat{\mu}(x, y) - \widehat{\mu}\right)^2 \tag{4.5}$$

is an estimator of the genetic variance in (2.11).

To estimate the environmental variance defined in (2.12), let

$$\sigma^2(x, y) = E\left[(W(x, y) - \mu(x, y))^2 |(x, y)\right] \tag{4.6}$$

for all genotypes $(x, y) \in \mathbb{G}$. Then,

$$\widehat{\sigma}^2(x, y) = \frac{1}{n(x, y) - 1} \sum_{\nu=1}^{n(x,y)} \left(W(x, y) - \widehat{\mu}(x, y)\right)^2 \tag{4.7}$$

is a conditionally unbiased estimator of $\sigma^2(x, y)$, given the genotype (x, y). Therefore,

$$\widehat{\text{var}}_E[W] = \sum_{(x,y)} \widehat{p}(x, y)\widehat{\sigma}^2(x, y) \tag{4.8}$$

is an estimator of the environmental variance defined in (2.12). From (2.13), it follows that an estimator of the phenotypic variance may be obtained by adding the estimators in (4.5) and (4.8), or this variance component could be estimated directly.

Given the estimators $\widehat{\mu}(x, y)$ for all genotypes in the sample, it would be straightforward to derive estimators of the three effects in the column vector

$\mathbf{\Phi}(x,j)$ in (3.24) for all genotypes (x,y) in the sample. Let $\widehat{\mathbf{\Phi}}(i,j)$ denote the estimator of the vector $\mathbf{\Phi}(x,j)$ for all genotypes under consideration. Then, let

$$\widehat{\mathbf{\Psi}}(x,y) = \widehat{\mathbf{\Phi}}(x,y)\widehat{\mathbf{\Phi}}^T(x,y) \tag{4.9}$$

denote an estimator of the matrix $\mathbf{\Psi}(x,y)$ in (3.26) for all genotypes (x,y). Given these definitions of estimators, it follows that

$$\widehat{\mathbf{\Psi}}_G = \sum_{(x,y)} \widehat{p}(x,y)\widehat{\mathbf{\Psi}}(x,y) \tag{4.10}$$

is an estimator of the genetic covariance matrix defined in (3.27). It should also be noted that an investigator would be free to estimate each component of the matrix $\widehat{\mathbf{\Psi}}_G$ separately.

It is also possible to estimate H_G, the measure of heritability defined in Section 2. From (2.13), it follows that

$$\widehat{\operatorname{var}}_P[W] = \widehat{\operatorname{var}}_G[W] + \widehat{\operatorname{var}}_E[W] \tag{4.11}$$

is an estimator of the phenotypic variance. Therefore,

$$\widehat{H}_G = \frac{\widehat{\operatorname{var}}_G[W]}{\widehat{\operatorname{var}}_P[W]} \tag{4.12}$$

is an estimator of H_G, a measure of heritability.

At any step in the development of software to implement the ideas under discussion, one could proceed in a number of directions. Suppose, for example, that an investigator was not inclined to estimate the matrix $\widehat{\mathbf{\Psi}}_G$ in (4.10). An alternative approach would be that of considering a remainder estimate $\widehat{R}_G$ which is defined by the equation

$$\widehat{\operatorname{var}}_G[W] = \widehat{\operatorname{var}}_G(A) + \widehat{\operatorname{var}}_G(\mathrm{IAI}) + \widehat{R}_G, \tag{4.13}$$

where $\widehat{\operatorname{var}}_G(A)$ and $\widehat{\operatorname{var}}_G(\mathrm{IAI})$ are estimates of the additive and intra-allelic-interactions variance components defined in (3.13) and (3.14). The remainder term $\widehat{R}_G$ would be a direct measure of the departure of the population from a Hardy–Weinberg equilibrium when equation (3.23) is valid. Observe that $\widehat{R}_G$ is the sum of all elements in the matrix $\widehat{\mathbf{\Psi}}_G$ off the principal diagonal.

If an investigator were interested in investigating whether the off-diagonal elements in this estimator of the covariance matrix would change significantly under the assumption that the population from which the sample was derived was in a Hardy–Weinberg equilibrium, the following procedure could be executed. Let $\widehat{p}(x)$ be an estimator of the marginal frequency of maternal alleles $x \in \mathbb{A}$ in the sample, and let the marginal frequency $\widehat{p}(y)$ be defined similarly for paternal alleles $y \in \mathbb{A}$ in the sample. Then, the next step in a computer simulation experiment with a goal of recomputing the estimate of the matrix $\widehat{\boldsymbol{\Psi}}_G$, under the assumption that the population was in a Hardy–Weinberg equilibrium, would be that of computing the product

$$p^*_{HW}(x, y) = \widehat{p}(x)\widehat{p}(y) \tag{4.14}$$

for all genotypes $(x, y) \in \mathbb{G}$ in the sample. Given this trial set of genotypic frequencies, the calculation procedures outlined above could be used to compute an alternative estimate of the covariance matrix $\boldsymbol{\Psi}_G$, symbolized by $\boldsymbol{\Psi}_{GHW}$, under the assumption that the population was in a Hardy–Weinberg equilibrium so that one would expect that the remainder term $\widehat{R}_G$ would be zero.

The direct method of estimation described above has many advantages when compared with classical methods of estimating variance components, because the effects defined in Section 3 may also be estimated directly from the data. By way of illustrative example, the direct estimator of the conditional expectation $\mu(y)$ is

$$\widehat{\mu}(x) = \sum_{y} \widehat{p}(y|x)\widehat{\mu}(x, y), \tag{4.15}$$

where

$$\widehat{p}(y|x) = \frac{\widehat{p}(x, y)}{\widehat{p}(x)} \tag{4.16}$$

for $\widehat{p}(x) \neq 0$. Therefore, the direct estimator of the additive effect defined in (3.9) is

$$\widehat{\alpha}(x) = \widehat{\mu}(x) - \widehat{\mu} \tag{4.17}$$

for all alleles $x \in \mathbb{A}$. A formula for the direct estimator of the effect $\alpha(y)$ is analogous to that of $\widehat{\alpha}(x)$. Given the estimators $\widehat{\alpha}(x)$ and $\widehat{\alpha}(y)$, a direct estimator of the measure of interaction between alleles x and y defined in (3.10) is

$$\widehat{\alpha}(x, y) = \widehat{\mu}(x, y) - \widehat{\mu} - \widehat{\alpha}(x) - \widehat{\alpha}(y) \tag{4.18}$$

for all genotypes $(x, y) \in \mathbb{G}$.

As can be seen from (3.13) and (3.14), the squares of the estimators of the effects defined above would be terms in the estimators of the additive and intra-allelic interaction components of variance so that if attention was focused only on the estimates of these variance components, an investigator may miss detecting the largest of the squared effects, which would be of interest in their own right. It is recommended, therefore, that the squares in the sets

$$\left\{\widehat{\alpha}^2(x), \widehat{\alpha}^2(y) | x \in \mathbb{A}, y \in \mathbb{A}\right\} \tag{4.19}$$

be calculated and inspected to get an idea as to which allele produces the largest additive effect. Similarly, it is recommended that the set of squares of measures of interaction

$$\left\{\widehat{\alpha}^2(x, y) | (x, y) \in \mathbb{G}\right\} \tag{4.20}$$

also be calculated and inspected to get an idea of which genotype has the largest measure of interaction of alleles.

It should also be mentioned that it would be desirable to work out the statistical properties of the estimators defined in this section. Included in these properties are consistency as sample sizes tend towards infinity and whether the estimators are unbiased. There is also a need for statistical tests to assess whether a particular estimate of a parameter was significantly different from zero. It is recognized that the working-out of these statistical properties would be important, but a full response to such statistical issues is beyond the scope of this paper. In this connection, it is interesting to note that computer intensive methods are now being used extensively in judging the statistical significance as to whether some region of a genome is implicated in some quantitative trait. For example, an interested reader may wish to consult the papers Raj *et al.* (2012) [19] and Rossin *et al.* (2011) [20], in which permutation tests have been used in assessing the statistical significance of hypothesized protein and other networks. It should also be mentioned that such computer-intensive methods as jackknifing and boot-strapping could also be used to assess the statistical significance of an estimate of an effect or variance component.

5. The Case of Two Autosomal Loci

Let $\mathbb{A}_1$ and $\mathbb{A}_2$ denote the set of alleles at locus 1 and 2, respectively. It will be assumed that each of these sets contains at least two alleles. In a diploid species with two sexes, such as humans, at every locus there is an allele contributed by the female parent and another allele contributed by the male parent. For the case of two autosomal loci, a genotype will be represented by the symbol (x_1, y_1, x_2, y_2), where (x_1, x_2) denotes the maternal alleles at the two loci and (y_1, y_2) are the corresponding paternal alleles. The set $\mathbb{G}$ all genotypes with respect to the two loci under consideration is the product set

$$\mathbb{G} = \mathbb{A}_1 \times \mathbb{A}_1 \times \mathbb{A}_2 \times \mathbb{A}_2. \tag{5.1}$$

To lighten the notation in what follows, let the vector $z = (x_1, y_1, x_2, y_2)$ denote a genotype $z \in \mathbb{G}$, and let $p(z)$ denote the frequency of genotype $z \in \mathbb{G}$ in the population. For some quantitative trait or character under consideration, let W denote a random variable describing the phenotypic variation with respect to some quantitative measurement among the individuals in a population. Then, given some genotype $z \in \mathbb{G}$, let the conditional expectation

$$\mu(z) = E\left[W|z\right] \tag{5.2}$$

denote the genetic value for this genotype. Just as in the case of one locus, this conditional expectation will play a basic role in defining measurements of the effects of each allele as well as the interactions among the two loci under consideration.

In general, one would not expect that the population under consideration would be in linkage equilibrium; consequently, it will be necessary to define a number of marginal and conditional distributions, which will be derived using the set

$$\mathfrak{D}_G = \left\{p(z)|z \in \mathbb{G}\right\} \tag{5.3}$$

of genotypic frequencies, which from now on will be called the genotypic distribution. For example, for allele x_1 suppose that we wish to derive a formula for the conditional expectation of $\mu(x_1, y_1, x_2, y_2)$, given x_1 with respect to the genotypic distribution. A first step in this derivation would be to calculate the marginal distribution

$$p(x_1) = \sum_{(y_1, x_2, y_2)} p(x_1, y_1, x_2, y_2) \tag{5.4}$$

for all $x_1 \in \mathbb{A}_1$. By definition, the conditional distribution of $\mu(x_1, y_1, x_2, y_2)$, given x_1, is

$$p(y_1, x_2, y_2 | x_1) = \frac{p(x_1, y_1, x_2, y_2)}{p(x_1)} \tag{5.5}$$

for $p(x_1) \neq 0$. Let $\mu(x_1)$ denote the conditional expectation of $\mu(x_1, y_1, x_2, y_2)$, given x_1. Then, by definition,

$$\mu(x_1) = \sum_{(y_1, x_2, y_2)} p(y_1, x_2, y_2 | x_1)\, \mu(x_1, y_1, x_2, y_2) \tag{5.6}$$

for all $x_1 \in \mathbb{A}_1$. The unconditional expectation of the $\mu(z)$ with respect to the genotypic distribution $\mathfrak{D}_G$, as expressed in a more succinct notation, is

$$\mu = \sum_{z \in \mathbb{G}} p(z)\mu(z). \tag{5.7}$$

Therefore, in analogy with the case of one allele, the additive effect of allele x_1 in the population will be defined as

$$\alpha(x_1) = \mu(x_1) - \mu \tag{5.8}$$

for all $x_1 \in \mathbb{A}_1$.

An analogous effect could be defined for each of the alleles y_1, x_2 and y_2, by applying the methods described for defining $\alpha(x_1)$. But, as will be demonstrated, for the case of two autosomal loci there are many more interaction terms that need to be defined. For example, for the case of a diploid species, there are four positions to be considered when classifying and defining effects and interactions among alleles. Consider, for example, the set of four alleles in each genotype $z = (x_1, y_1, x_2, y_2) \in \mathbb{G}$, and let $\mathfrak{S} = \{1, 2, 3, 4\}$ denote the set of four positions that need to be considered with respect to two loci with two alleles at each locus that were contributed by the maternal and the paternal parent, respectively. To provide a framework for describing various types of interactions among the alleles at the two loci under consideration, it will be helpful to consider the class of all subsets of the four positions. Let $\mathfrak{T}$ denote the class of all subsets of $\mathfrak{S}$. Included in the class $\mathfrak{T}$ is the empty set φ as well as subsets containing 1, 2, 3 and 4 elements of the set $\mathfrak{S}$. As is well-known from combinatorial analysis, the total number of sets in $\mathfrak{T}$ is $2^4 = 16$, and, as is also well known from combinatorics, the equation

$$\binom{4}{0} + \binom{4}{1} + \binom{4}{2} + \binom{4}{3} + \binom{4}{4} = 2^4 = 16 \tag{5.9}$$

is valid. For $\nu = 0, 1, 2, 3, 4$, let $\mathfrak{T}_\nu$ denote the subclass of sets in $\mathfrak{T}$ that contain ν elements. Then, as can be seen from equation (5.9), each of the subclasses $\mathfrak{T}_0$ and $\mathfrak{T}_4$ contains one set, namely φ and $\mathfrak{S}$, respectively. Similarly, each of the subclasses $\mathfrak{T}_1$ and $\mathfrak{T}_3$ contains four sets, and the subclass $\mathfrak{T}_2$ contains six sets. Recall that

$$\binom{4}{2} = 6. \tag{5.10}$$

To describe a framework in which to quantify the ideas of intra-allelic interactions and epistatic interactions among alleles at different loci, it will be helpful to enumerate the sets in the subclasses $\mathfrak{T}_1$, $\mathfrak{T}_2$ and $\mathfrak{T}_3$ in terms of elements of the set $\mathfrak{S}$. For example,

$$\mathfrak{T}_1 = (\{1\}, \{2\}, \{3\}, \{4\}) \tag{5.11}$$

is the class of singletons, which are subsets that contain only one element of $\mathfrak{S}$. It is this subclass of sets that was used to define the additive effects mentioned above. The subclass $\mathfrak{T}_2$ of sets has the explicit form

$$\mathfrak{T}_2 = (\{1, 2\}, \{1, 3\}, \{1, 4\}, \{2, 3\}, \{2, 4\}, \{3, 4\}). \tag{5.12}$$

At this point recall that positions 1 and 2 in the set $\mathfrak{S}$ are those for the two alleles at locus 1, and positions 3 and 4 in this set are those for the two alleles at locus 2. Therefore, the two sets of positions in subclass

$$\mathfrak{T}_{2\mathrm{IAI}} = (\{1, 2\}, \{3, 4\}) \tag{5.13}$$

will be used to define effects that measure intra-allelic interactions at the two loci under consideration. On the other hand, the pairs of positions in the subclass

$$\mathfrak{T}_{2\mathrm{EPI}} = (\{1, 3\}, \{1, 4\}, \{2, 3\}, \{2, 4\}) \tag{5.14}$$

represent positions from different loci. Consequently, sets in this class will form a basis for defining effects that measure epistatic interactions among the alleles at the two loci under consideration. The subsets in the subclass $\mathfrak{T}_3$ are as follows:

$$\mathfrak{T}_3 = (\{1,2,3\},\{1,2,4\},\{1,3,4\},\{2,3,4\}). \tag{5.15}$$

The sets in this class form a basis for defining effects that measure the effect that an allele at one locus may affect or modify intra-allelic interactions at another locus. For example, the two sets in the subclass

$$\mathfrak{T}_{31\mathrm{EPI}} = (\{1,2,3\},\{1,2,4\}) \tag{5.16}$$

would form a basis for defining an effect measuring intra-allelic interactions at locus 1 that may affect the expression of alleles in positions 3 and 4 at locus 2. Similarly, the sets in the subclass

$$\mathfrak{T}_{32\mathrm{EPI}} = (\{1,3,4\},\{2,3,4\}) \tag{5.17}$$

would form a basis for defining an effect measuring intra-allelic interactions at locus 2 that may be affected by alleles at positions 1 and 2 at locus 1.

For cases in which many alleles can be recognized at each locus, it would be necessary to develop a nomenclature to describe many types of interactions among the alleles at the two autosomal loci under consideration, as will be illustrated below. In this connection, an interested reader may wish to consult the pioneering work of Cockerham (1954) [5], which describes a nomenclature for various epistatic effects and components of the genetic variance. For example, effects and variance components corresponding to the sets in the class $\mathfrak{T}_{2\mathrm{IAI}}$ would be labeled dominant for either the effects or variance components and would be denoted by the symbol D, whereas those in the class $\mathfrak{T}_{2\mathrm{EPI}}$ would be labeled additive by additive effects or variance components and denoted by the symbol AA. One could proceed in this way to develop a nomenclature of the 15 effects and variance components under consideration. But this type of nomenclature will not be used in this paper and epistasis will be described in terms of sets and effects as well as variance components.

The first step in defining these effects is to derive a formula for the conditional expectation of a genetic value $\mu(z)$, given every set A of positions such that

$$A \in E = \bigcup_{\nu=1}^{3} \mathfrak{T}_{\nu}. \tag{5.18}$$

To define these effects, it will be helpful to introduce a succinct notation. For every set A of positions, let A^c denote the complement of this set with respect to the set $\mathfrak{S}$, and let $z(A)$ and $z(A^c)$ denote subsets of alleles in z corresponding to the positions in the sets A and A^c, respectively. In what follows, the symbol $z(A), z(A^c)$ will stand for the union of the positions in the two sets. Given this notation, the marginal distribution $p(z(A))$ is defined by

$$p(z(A)) = \sum_{z(A^c)} p(z(A), z(A^c)) \tag{5.19}$$

for every $z(A) \in \mathbb{G}(A)$, where $\mathbb{G}(A)$ is a subset of $\mathbb{G}$ containing only those alleles corresponding to the positions in the set A. Thus, in this succinct notation,

$$p(z(A^c)|z(A)) = \frac{p(z(A), z(A^c))}{p(z(A))} \tag{5.20}$$

is the conditional distribution of $z(A^c)$, given $z(A)$ for $p(z(A)) \neq 0$. Let $\mu(z(A))$ denote the conditional expectation of $\mu(z)$, given $z(A)$. Then,

$$\mu(z(A)) = \sum_{z(A^c)} p(z(A^c)|z(A))\mu(z(A), z(A^c)) \tag{5.21}$$

for every $A \in E$.

Given formula (5.21), one may proceed systematically through each of the sets in the union E in (5.18) to calculate $\mu(z(A))$ for every $A \in E$. For example, suppose that $A = \{1\}$. Then, $\mu(z(A)) = \mu(x_1)$ for all $x_1 \in \mathbb{A}_1$. By continuing in this manner, all the conditional pairs of expectations, $(\mu(x_\nu), \mu(y_\nu))$, for $\nu = 1, 2$, could be computed, and formula (5.8) could be used to compute the four effects: $\alpha(x_\nu), \alpha(y_\nu)$ for $\nu = 1, 2$.

Similarly, for every set $A \in \mathfrak{T}_2$, $\mu(z(A))$ would need to be calculated. Suppose, for example, that $A = \{1, 2\}$. Then, $\mu(x_1, y_1)$ would need to be calculated for every $(x_1, y_1) \in \mathbb{A}_1 \times \mathbb{A}_1$. Then, as in the case of one locus, the intra-allelic effect $\alpha(x_1, y_1)$ would be defined by

$$\alpha(x_1, y_1) = \mu(x_1, y_1) - \mu - \alpha(x_1) - \alpha(y_1). \tag{5.22}$$

By continuing in this way, an effect $\alpha(z(A))$ could be defined for every subset $A \in \mathfrak{T}_3$. To illustrate how each of these four effects could be defined, consider the case $A = \{1, 2, 3\}$. In this case, $\mu(z(A)) = \mu(x_1, y_1, x_2)$ for all $(x_1, y_1, x_2) \in \mathbb{A}_1 \times \mathbb{A}_1 \times \mathbb{A}_2$. Then, by definition, the effect $\alpha(x_1, y_1, x_2)$ is

$$\begin{aligned}\alpha(x_1, y_1, x_2) &= \mu(x_1, y_1, x_2) - \mu - \alpha(x_1) - \alpha(y_1) - \alpha(x_2)\\ &\quad -\alpha(x_1, y_1) - \alpha(x_1, x_2) - \alpha(y_1, x_2).\end{aligned} \tag{5.23}$$

Altogether, for the subclass $\mathfrak{T}_3$, four effects would need to be computed, using the procedure illustrated in (5.23). Note that all the effects on the right in this equation, were defined for each subset of the set of symbols $\{x_1, y_1, x_2\}$. This procedure may also be used to set down formulas for each of the three remaining subsets in the subclass $\mathfrak{T}_3$. Furthermore, in formulations in which more than two loci were under consideration, the procedure (5.23) used to define the effects for the case of two loci could be extended to defining effects for some number of loci $n \geq 3$. The last step in defining effects for the two-locus case is to define the effect $\alpha(z(\mathfrak{S})) = \alpha(x_1, y_1, x_2, y_2)$ for all genotypes $z \in \mathbb{G}$. In this connection let $\alpha(z(\mathfrak{S})) = \alpha(z)$ be such that the equation

$$\mu(z) = \mu + \sum_{A \in \mathfrak{T}_1} \alpha(z(A)) + \sum_{A \in \mathfrak{T}_2} \alpha(z(A)) + \sum_{A \in \mathfrak{T}_3} \alpha(z(A)) + \alpha(z) \tag{5.24}$$

holds for all genotypes $z \in \mathbb{G}$.

Having defined the set of 15 effects for the case of two autosomal loci, the next step is that of defining components of the genetic variance. For example, the additive genetic variance is defined by

$$\text{var}_A[W] = \sum_{A \in \mathfrak{T}_1} E_{\mathfrak{D}_G}[\alpha^2(z(A))], \tag{5.25}$$

where the expectation is taken with respect to the genotypic distribution $\mathfrak{D}_G$. The intra-allelic interaction component of the genetic variance is defined by

$$\text{var}_{IAI}[W] = \sum_{A \in \mathfrak{T}_{2IAI}} E_{\mathfrak{D}_G}[\alpha^2(z(A))], \tag{5.26}$$

and the epistatic component of the genetic variance with respect to two loci is defined by

$$\mathrm{var}_{EPI}[W] = \sum_{A \in \mathfrak{T}_{2EPI}} E_{\mathfrak{D}_G}[\alpha^2(z(A))]. \tag{5.27}$$

For the case of three alleles, the equation

$$\mathrm{var}_{IAI1}[W] = \sum_{A \in \mathfrak{T}_{31EPI}} E_{\mathfrak{D}_G}[\alpha^2(z(A))]$$

is the component of variance for intra-allelic interaction at the first locus that may be modified by an allele at the second locus. Similarly, the component of the genetic variance for intra-allelic interaction at the second locus that may be modified by an allele at the first locus is

$$\mathrm{var}_{\mathrm{IAI2}}[W] = \sum_{A \in \mathfrak{T}_{32\mathrm{EPI}}} E_{\mathfrak{D}_G}[\alpha^2(z(A))]. \tag{5.28}$$

Finally, the component of the genetic epistatic variance as measured by effects $\alpha(z)$ is defined by

$$\mathrm{var}_{\mathrm{EPI4}}[W] = \sum_{A \in \mathfrak{T}_4} E_{\mathfrak{D}_G}[\alpha^2(z(A))]. \tag{5.29}$$

It should be noted that the set of components of the genetic variance was defined arbitrarily, but a user of the ideas presented in this section may wish to adapt another nomenclature for the set of 15 effects and components of the total genetic variance.

An experimenter could test whether a sample of individuals whose genotypes had been determined with respect to two autosomal loci was in linkage equilibrium, but in any case it would be of interest to compute the genetic covariance matrix for the case under consideration. Let A denote any set of positions in the union

$$A \in \mathfrak{A} = \bigcup_{\nu=1}^{4} \mathfrak{F}_\nu \tag{5.30}$$

and let

$$\mathbf{\Phi}(z) = (\alpha(z(A)) | A \in \mathfrak{A}) \tag{5.31}$$

denote a 15×1 vector of classes of effects. Observe that within each class of effects corresponding to a set A there would be a collection of effects corresponding to the number of alleles at each locus. A useful ordering of the effects in this vector would be to let the subset of singletons be the first four elements of the vector, the six sets of pairs of positions would be the next six elements in the vector, the next four elements of the vector would be the four effects corresponding to the sets of triples of positions, and lastly the effect for the singleton $\mathfrak{S}$ would be the last 15th effect in the column vector. As was tacitly used in the definitions of the components of the genetic variance listed above, each effect has the unconditional expectation

$$E_{\mathfrak{D}_G}[\alpha(A(z))] = 0 \tag{5.32}$$

for all $A \in \mathfrak{A}$. Let

$$\Psi(z) = \Phi(z)\Phi^T(z)$$

denote a 15×15 matrix of products of effects for the genotypes $z \in \mathbb{G}$. Then, by definition, the covariance matrix of the vector $\Phi(z)$ of effects is

$$\Psi_G = \sum_{z \in \mathbb{G}} p(z)\Phi(z)\Phi^T(z) = E_{\mathfrak{D}_G}[\Psi(z)]. \tag{5.33}$$

As part of an analysis of data, at this point in the calculations, a data annalist may wish to compute the eigen values and vectors of the symmetric matrix Ψ_G. It would also be of interest to inspect the off-diagonal components of the matrix Ψ_G to provide an assessment of the impact of effects on the components of the genetic variance when the population is not in linkage equilibrium at the two loci under consideration. On the other hand, an investigator may not wish to compute and analyze the matrix Ψ_G in (5.33) and would be content with an estimate of the fraction

$$\frac{\text{var}_{\text{EPI}}[W] + \text{var}_{IAI2}[W] + \text{var}_{\text{EPI4}}[W]}{\text{var}_G[W]}, \tag{5.34}$$

where $\text{var}_G[W]$ is the total genetic variance. An estimate of this ratio would be of interest, because it would provide an investigator with some idea of the significance of the contribution of epistatic effects to the total genetic variance. At the same time, it should be recognized that the formula (5.34) was derived under the assumption that the population was not in linkage equilibrium and could be biased by negative covariance terms.

An investigator may, therefore, also wish to carry out a computer simulation experiment under the assumption that the population or sample was in linkage equilibrium. The first step in setting up such a computer experiment would be that of computing the marginal allele probabilities. Let $p_1(x_1)$ and $p_1(y_1)$ denote the marginal probabilities, respectively, for the maternal and paternal alleles at locus 1, and define the marginal $p_2(x_2)$ and $p_2(y_2)$ for locus 2 similarly. Then, the simulated population would be in linkage equilibrium if the genotypic probabilities $p(z)$ satisfied the equation $p(z) = p_1(x_1)p_1(y_1)p_2(x_2)p_2(y_2)$ for all genotypes $z = (x_1, y_1, x_2, y_2) \in \mathbb{G}$. Given these assigned genotypic probabilities, an investigator could carry out a computer simulation experiment under the assumption that the sample or population was in linkage equilibrium.

Just as in the one-locus case considered in Section 4, it is recommended that an investigator inspect the squares of all effects defined above for the case of two autosomal loci. For example, the set of squares of additive effects is defined by

$$\mathfrak{E}_1 = \left\{\alpha^2(z(A))|A \in \mathfrak{T}_1\right\}. \tag{5.35}$$

It will be tacitly assumed that the elements in the set $\mathfrak{E}_1$ are estimates of effects so as to simplify the notation. For the case where each locus has two alleles, the set $\mathfrak{E}_1$ would contain a small number of elements so that an investigator could easily find the largest one. Similarly, the set of squared effects that are measures of intra-allelic interactions is defined by

$$\mathfrak{E}_{2IAI} = \left\{\alpha^2(z(A))|A \in \mathfrak{T}_{2IAI}\right\}. \tag{5.36}$$

Like the set $\mathfrak{E}_1$ for the case of two alleles at each of the two loci under consideration, the set $\mathfrak{E}_{2IAI}$ would contain a small number of estimated squared effects so that an investigator could easily find the largest one. By continuing this way, the set of estimated squared effects corresponding to each of the subclasses of effects defined above for various types of epistatsis could also be defined but the enumeration of these sets will be left as an exercise for the interested reader.

In an interesting paper, Hemani *et al.* (2013) [8] an evolutionary perspective on epistasis and the missing heritability was the focus of attention. These authors assert that results of genome wide association studies may be improved if epistatic effects may be searched for explicitly. It is suggested that the epistatic effects defined in this section may also be useful in genome-wide association studies.

6. An Overview of the Case of 11 Autosomal Loci

As mentioned in the introduction, there is an interesting and important case in human genetics pertaining to Alzheimer's disease (AD) in which there is a developing consensus that 11 autosomal regions, loci, of the human genome have been implicated in this disease. It is suggested that the interested reader may wish to consult the paper by Raj *et al.* (2012) [19] and the literature cited therein for more details regarding these genomic regions. In studies of patients with AD, quantitative measurements are often made on each patient so that AD may be viewed as a quantitative trait in humans. It is, therefore, of interest to provide an overview of an extension of the structure for the case of two autosomal loci developed in Section 5 to the case of 11 autosomal loci.

For a diploid species such as humans, two alleles occupy each locus so that for the case of 11 loci, there are $11 \times 2 = 22$ positions to consider in the set

$$\mathfrak{S} = (s \mid s = 1, 2, \ldots, 22) \tag{6.1}$$

of positions. Therefore, in this case, the class $\mathfrak{T}$ of all subsets of $\mathfrak{S}$ contains

$$2^{22} = 4{,}194{,}304 \tag{6.2}$$

sets. Included in $\mathfrak{T}$ is the empty set φ so, just as for the case of two loci, no effect will be associated with φ. It follows, therefore, that in theory, $2^{22} - 1 = 4{,}194{,}303$ effects could be defined for the case of 11 autosomal loci, but it is unlikely that any investigator would attempt to estimate such a large number of effects.

When one is dealing with 11 or more autosomal loci, it is also important to remember that for the case of many loci, one should keep in mind the caveat that the number of possible genotypes under consideration may be quite large and exceed the sample size that is available to an investigator or investigators. For the case of 11 autosomal loci and two alleles per locus, each vector in the pair (x, y) denoting a genotype would contain 11 alleles contributed by the maternal and the paternal parent, respectively. Thus, if it were possible to determine the parental source of each allele, one could in principle identify four genotypes per locus. Therefore, if 11 autosomal loci were under consideration, the number of genotypes that could be identified would be

$$4^{11} = 4{,}194{,}304. \tag{6.3}$$

Observe that this is the same number as that in (6.2), and, moreover, it in all likelihood exceeds the number of individuals in any sample of individual whose genomes have been sequenced that are presently available to investigators.

Consider, for example, the case of 11 autosomal loci with two alleles at each locus and suppose that an investigator identifies three genotypes per locus; namely two homozygotes and one heterozygote at each locus. In such circumstances, an investigator may not be able to determine whether any allele was contributed by the maternal or paternal parent. Under this assumption that only three genotypes can be identified per locus, it follows that the total number of "genotypes" that could be identified with respect to 11 autosomal loci would be

$$3^{11} = 177{,}147. \tag{6.4}$$

A number of this magnitude would in all likelihood exceed the sample size available to present day investigators, particularly if it is required that all individuals in the sample have had their genomes sequenced. If a sample size is considerably smaller than the number in (6.4), then it is recommended that an investigator confine attention to some subset $\mathfrak{S}_1$ of loci and individuals in a sample such that for each identifiable genotype (x, y), the number of individuals, $n(x, y) \geq 1$, with this genotype is sufficiently large so that one may make reliable and statistically significant genetic inferences based on the available data.

For the case of 11 autosomal loci, a sample available to an investigator may not be sufficiently large to accommodate the set of possible genotypes, because the number of individuals of all genotypes may not be sufficiently large for one to draw reliable statistical inferences. However, when attention is focused on a subset of loci, the number of individuals for each genotype with respect to this subset of loci is sufficiently large for one to draw reliable statistical inferences. By way of an illustrative and hypothetical example, suppose that an investigator was able to find a sufficient sample size for each genotype with respect to six autosomal loci with three distinguishable genotypes at each locus. Let $\mathbb{G}_S$ denote the set of genotypes in the sample and let $n(x, y)$ denote the number of individuals in the sample of genotype $(x, y) \in \mathbb{G}_S$. For the case of six autosomal loci, the total number of effects that may be defined is

$$2^{12} - 1 = 4{,}095. \tag{6.5}$$

It is doubtful that any investigation would have the persistence or interest to estimate 4,095 effects, but it may be of interest to estimate only first, second and third order effects. It is easy to see that the number of first order or additive effects would be

$$\binom{12}{1} = 12, \tag{6.6}$$

and as in (5.11) let $\mathfrak{T}_1$ denote the class of subsets of the set $\mathfrak{S} = (1, 2, \ldots, 12)$ of positions containing one position. It is straightforward to enumerate the sets in the class $\mathfrak{T}_1$. For the case of 12 positions, the number of subsets of $\mathfrak{S}$ containing two positions is

$$\binom{12}{2} = 66. \tag{6.7}$$

Let $\mathfrak{T}_2$ denote the class of subsets of $\mathfrak{S}$ containing two positions. Similarly, the number of subsets of $\mathfrak{S}$ containing three positions is

$$\binom{12}{3} = 220. \tag{6.8}$$

Observe that if an investor chose to follow the procedure just outlined, the total number of effects that would need to be defined would be

$$12 + 66 + 220 = 298. \tag{6.9}$$

Let $\mathfrak{T}_3$ denote the class of subsets of $\mathfrak{S}$ containing three positions.

It is interesting to note that the enumeration of the sets in the classes $\mathfrak{T}_2$ and $\mathfrak{T}_3$ may be accomplished by using a type of recursive procedure. To describe this recursive procedure, it is helpful if the notation is extended to include the number of loci and positions under consideration. For example, let $\mathfrak{T}_2^{(\nu)}$ denote a class of subsets of two positions taken from the sets of positions $\mathfrak{S}_\nu$ for $\nu = 4, 6, \ldots, 12$ sets of positions corresponding to $l = 2, 3, \ldots, 6$ loci. Then, it follows that the containment relations

$$\mathfrak{T}_2^{(4)} \subset \mathfrak{T}_2^{(6)} \subset \mathfrak{T}_2^{(8)} \subset \mathfrak{T}_2^{(10)} \subset \mathfrak{T}_2^{(12)} \tag{6.10}$$

hold. Thus, if an investigator has enumerated the subsets in the class $\mathfrak{T}_2^{(4)}$ for the case of two loci (see Section 5), then to extend this enumeration to the case of three loci and six positions, one could add positions 5 and 6 to the set $\mathfrak{S}_4$ to obtain the set $\mathfrak{S}_6$ of positions for the case of three loci. The next step in this recursive process would be that of adding to $\mathfrak{T}_2^{(4)}$ those sets with two positions that include position 5 and 6 to obtain all the subsets with two positions from the set $\mathfrak{S}_6$. By continuing in this recursive manner, the set of two positions in the class $\mathfrak{T}_2^{(12)}$ could be enumerated. It is also of interest to note that the containment relations

$$\mathfrak{T}_3^{(4)} \subset \mathfrak{T}_3^{(6)} \subset \mathfrak{T}_3^{(8)} \subset \mathfrak{T}_3^{(10)} \subset \mathfrak{T}_3^{(12)} \tag{6.11}$$

for classes of subsets containing sets with three positions are also valid. Therefore, the class of sets $\mathfrak{T}_3^{(12)}$ could also be enumerated by using a recursive procedure. It is also highly plausible that a clever computer programmer could write code to accomplish the enumeration of the classes of sets $\mathfrak{T}_2^{(12)}$ and $\mathfrak{T}_3^{(12)}$.

Given the enumerated classes of subsets $\mathfrak{T}_1, \mathfrak{T}_2$ and $\mathfrak{T}_3$, the next step in providing an overview of the case of six autosomal loci is that of defining an effect for each set in the three classes of subsets. Briefly, the procedures used in defining and setting up algorithms to compute them are given implicitly in equations (5.19), (5.20) and (5.21). Let

$$\mathfrak{C}_1 = \{\alpha(z(A)) | A \in \mathfrak{T}_1\} \tag{6.12}$$

denote the set of first order effects. Similarly, let

$$\mathfrak{C}_2 = \{\alpha(z(A)) | A \in \mathfrak{T}_2\}, \tag{6.13}$$

$$\mathfrak{C}_3 = \{\alpha(z(A)) | A \in \mathfrak{T}_2\} \tag{6.14}$$

denote, respectively, the class of second and third order effects. It should be noted that the formulas for computing the first, second and third order effects are outlined in formulas (5.22) and (5.23) for each combination of alleles.

Just as suggested for the case of two autosomal loci in Section 5, it would be of interest to find the largest of the squares of each effect to get some idea as to which effect contributes the most to a variance component under consideration. An explicit form of the squares of first order effects is

$$\mathfrak{D}_1 = \left\{\alpha^2(\nu) | \nu \in \mathfrak{S}\right\}, \tag{6.15}$$

where $\mathfrak{S} = (s | s = 1, 2, \ldots, 12)$. For the sake of simplicity, suppose that there are only two alleles at each the the six loci under consideration. Under this assumption, each of the 12 positions may be occupied by either of the two alleles at each locus. Therefore, the number of squared values in the set $\mathfrak{D}_2$ is 24. Let $\mathfrak{D}_2$ and $\mathfrak{D}_3$ denote, respectively, the set of squared effects from the sets $\mathfrak{C}_2$ and $\mathfrak{C}_3$. Suffice it to say that for the case of two alleles at each of the six loci, the number of squared effects in each of the sets $\mathfrak{D}_\nu$ for $\nu = 1, 2, 3$ could be determined, but this exercise will be left to the interested reader.

If an investigator does not have a sufficiently large sample to work with for the case of six autosomal loci, then a reduced version of the ideas just outlined could be used to study a smaller number of loci such that the sample size for each genotype is sufficiently large to draw statistically reliable conclusions. Given the ideas just outlined, a study of cases for two, three or four loci may be feasible if there is insufficient data for studying the case of five or six autosomal loci. It should be noted that for the case where only three genotypes per locus may be identified, the number of effects that an investigator could estimate would be significantly smaller than for the case where four genotypes may be identified per locus. It is beyond the scope of this paper to consider the case of only three identifiable genotypes per locus, but the details for this case will be worked out in subsequent papers for one or more autosomal loci.

When one considers the union of the sets $\mathfrak{D}_1, \mathfrak{D}_2$ and $\mathfrak{D}_3$, it is easy to see that many tests of statistical significance may need to be made if an investigator wishes to assess the statistical significance of some chosen number of squared effects. It is beyond the scope of this paper to deal with the problem of making many statistical tests and computing measures of statistical significance, but it is suggested that the reader may wish to consult the literature on this subject. Included among the papers that would be interest to consult are Benjamini *et al.* (1995) [1], (2001) [2] and (2005) [3].

A version of equation (5.24) may also be set down for the case of six autosomal loci under consideration and has the form

$$\mu(z) = \mu + \sum_{A \in \mathfrak{T}_1} \alpha(z(A)) + \sum_{A \in \mathfrak{T}_2} \alpha(z(A)) + \sum_{A \in \mathfrak{T}_3} \alpha(z(A)) + \alpha_R(z) \tag{6.16}$$

for all genotypes $z \in \mathbb{G}_S$, where $\alpha_R(z)$ is a remainder effect. In principle, if all the effects on the right hand side of the equation have been estimated for all genotypes $z \in \mathbb{G}_S$, then the effect $\alpha_R(z)$ could be estimated for all genotypes $z \in \mathbb{G}_S$. Given these estimates, one could then proceed to estimate the variance component corresponding to the effect $\alpha_R(z)$, by using the formula

$$\text{var}_R[W] = \sum_{z \in \mathbb{G}} p(z)\alpha_R^2(z). \tag{6.17}$$

Let $\widehat{\text{var}}_G[W]$ denote an estimate of the genetic variance defined in equation (2.11) and let $\widehat{\text{var}}_R[W]$ denote an estimate of the variance component in (6.17). Then, the ratio

$$\frac{\widehat{\text{var}}_R[W]}{\widehat{\text{var}}_G[W]} \tag{6.18}$$

may be used as an estimate of the fraction of the total genetic variance that is attributable to the remainder effects $\alpha_R(z)$ for all genotypes $z \in \mathbb{G}_S$.

When interpreting this estimate, an investigator should also be aware of the possibility that the sample of individuals that constitute the data used to estimate all effects and variance components may not be in linkage equilibrium with respect to the six autosomal loci under consideration. In this case, it may be worthwhile to compute a version of the genetic covariance $\boldsymbol{\Psi}_G$ defined in (3.27) for the case of six loci. It can be shown that in terms of this matrix, the estimate $\widehat{\text{var}}_G[W]$ of the genetic variance may be represented in the form

$$\widehat{\text{var}}_G[W] = 1^T \widehat{\boldsymbol{\Psi}}_G 1, \tag{6.19}$$

where 1 is a column of 1s, T denotes the transpose of the vector or matrix and $\widehat{\boldsymbol{\Psi}}_G$ is an estimate of $\boldsymbol{\Psi}_G$. Given this matrix, all variance components associated with equation (6.16) would be on the principal diagonal of the matrix $\boldsymbol{\Psi}_G$. Therefore, the trace of the matrix, the sum of the elements on the principal diagonal of $\widehat{\Psi}_G$, is the sum of the variance components corresponding to the effects in equation (6.16). Let $\widehat{\text{trace}}[\widehat{\Psi}_G]$ denote an estimate of the sum of these variance components. Then, the ratio

$$\frac{\widehat{\text{trace}}\left[\widehat{\Psi}_G\right]}{\widehat{\text{var}}_G[W]} \tag{6.20}$$

is an estimate of the fraction of the total genetic variance that is attributable to the variance components defined in connection with equation (6.16). It would also be of interest to inspect the elements in the matrix $\widehat{\Psi}_G$ off the principal diagonal to make an assessment as to the effects that nonzero covariance terms contribute to the estimate of the total genetic variance in (6.19).

This ratio may be interpreted as a measure of the genetic variance that is attributable to the effects defined in connection with the construction of equation (6.16), taking into account that these effects may be correlated for the case where the sample of individuals is not in linkage equilibrium. If this ratio is equal to 1, then the variance components defined in connection with equation (6.16) are sufficient to account for all the genetic variance in the quantitative trait under consideration. But if this ratio is less than 1, then these components of the genetic variance would not be sufficient to account for the total genetic variance. It is also appropriate to mention that the ratio in (6.20) may be computed without computing the matrix $\widehat{\Psi}_G$, by computing each variance component corresponding to the effects in equation (6.16), using formulas analogous to (6.17).

It is recognized that an investigator who wishes to apply the ideas on the estimation of effects and variance components set forth in this paper may also want to test some statistical hypotheses, but it is beyond the scope of this paper to suggest various types of statistical tests of significance in addition to those mentioned briefly above.

ACKNOWLEDGMENTS

A word of thanks is due to Dr. Towfique Raj, Division of Genetics, Brigham and Women's Hospital, Harvard Medical School, Boston, MA 02115, USA, who called the author's attention to recent papers devoted to the estimation of heritability of various quantitative traits in humans, which have been cited in the paper. It should also be mentioned that a cooperative research effort involving the author and Dr. Raj's group is also in progress, with a goal of writing software to implement the ideas set forth in this paper, along with results not included in this paper, and applying them in a quantitative genetic analysis of data from samples of patients whose genomes have been sequenced.

REFERENCES

1. Benjamini, Y. and Hochberg, Y. (1995) Controlling false discovery rate: a practical and powerful approach to multiple testing. *J. R. Soc. Ser. B* **57**: 289–300.
2. Benjamini, Y. and Yekutieli, D. (2001) The control of false discovery rate under dependency. *Ann. Stat.* **29**: 1165–1188.
3. Benjamini, Y. and Yekutieli, D. (2005) False discovery rate adjusted multiple confidence intervals for selected parameters. *J. Am. Stat. Assoc.* **100**: 71–93.
4. Bulmer, M. G. (1980) *The Mathematical Theory of Quantitative Genetics.* Clarendon, Oxford.
5. Cockerham, C. C. (1954) An extension of the concept of partitioning the hereditary variance for analysis of covariance among relatives when epistasis is present. *Genetics* **39**: 859–882.
6. Falconer, D. and MacKay, T. F. C. (1996) *Introduction to Quantitative Genetics.* Longman, New York.
7. Fisher, R. A. (1918) The correlation among relatives on the assumption of Mendelian inheritance. *Trans. R. Soc. Edinb.* **52**: 399–433.
8. Hemani, G., Knott, S. and Haley, C. (2013) An evolutionary perspective on epistasis and the missing heritability. *PLoS Genet.* **9(2)**: e1003295. DOI: 10.1371/journal.pgen.1003295.
9. Kao, C.-H. and Zeng, Z.-B. (2002) Modeling epistasis of quantitative trait loci using Cockerham model. *Genetics* **160**: 1243–1261.
10. Kempthorne, O. (1954) The correlations between relatives in a random mating population. *Proc. R. Soc. London B* **143**: 103–113.
11. Kempthorne, O. (1957) *An Introduction to Genetic Statistics.* John Wiley and Sons, New York.
12. Laird, N. M. and Lange, C. (2011) *The Fundamentals of Modern Statistical Genetics.* DOI: 10.1007/978-1-4419-7338-2. Springer, New York, Dordrecht, Heidelberg, London.
13. Liu, B. H. (1998) *Statistical Genomics; Linkage, Mapping and QTL Analysis.* CRC, Boca Raton, London, New York and Washington, D.C.

14. Lynch, M. and Walsh, B. (1998) *Genetics and the Analysis of Quantitative Traits.* Sinauer Associates, Sunderland, MA, USA.
15. Mao, Y., Nicole, R., Ma, L., Dvorkin, D. and Da. Y. (2006) Detection of SNP epistasis effects of quantitative traits using extended Kempthorne model. *Physiol. Genomics* **28**: 46–52.
16. Mode, C. J. and Robinson, H. F. (1959) Pleiotropism and the genetic variance and covariance. *Biometrics* **15**: 518–537.
17. Mode, C. J. (2014) Estimating statistical measures of pleiotropic and epistatic effects in the genomic era. *Int. J. Stat. Probab.* **3**(2), 81–100.
18. Price, A. L., Helgason, A., *et al.* (2011) Single tissue and cross-tissue heritability of gene expression via identity-by-descent in related and unrelated individuals. *PloS Genet.* **7**(2): e1001317. DOI: 10.1371/journal.pgen.1001317.
19. Raj, T., Shulman, J. M., Keenan, B. T. Lori B. Chibnik, L. B., Evans, D. A., Bennett, D. A., Stranger, B. E. and De Jager, P. L. (2012) Alzheimer disease susceptibility loci: evidence for a protein network under natural selection. DOI: 10.1016/j.ajhg.2012.02.022. _2012. The American Society of Human Genetics.
20. Rossin, E. J., Lage K., Raychaudhuri, S., Xavier, R. J., Tatar, D., *et al.* (2011) Proteins encoded in genomic regions associated with immune-mediated disease physically interact and suggest underlying biology. *PLoS Genet.* **7**(1): e1001273. DOI: 10.1371/journal.pgen.1001273.
21. Stranger, B. E., Eli A., Stahl, E. A. and Raj. T. (2010) Progress and promise of genome-wide association studies for human complex trait genetics. DOI: 10.1534/genetics.110.120907.
22. Wu, R.-L., Ma, C.-X. and Casella, G. (2010) *Statistical Genetics of Quantitative Traits: Linkage, Maps and QTL.* Springer Science.
23. Yang, J., Benyamin, B., *et al.* (2010) Common SNPs explain a large proportion of heritability for human height. *Nat. Genet.* **42**(7): 565–569. DOI: 10.1038/ng.608.
24. Zaitlen, N., Kraft, P., *et al.* (2013) Using extended genealogy to estimate components of heritability for 23 quantitative and dichotomous traits. *PLoS Genet.* **9**(5): e1003520. DOI: 10.1371/journal.pgen.1003520.

CHAPTER 4

Estimating Statistical Measures of Pleiotropic and Epistatic Effects in the Genomic Era*

ABSTRACT

Recent developments in the technology for sequencing the genomes of various species have had a profound effect on the working paradigms of various fields of genetics. Included among these fields is the classical field of quantitative genetics, which is a subfield of statistical genetics that is devoted to traits that can be quantified on some continuous scale and are often influenced by alleles at many loci. In recent years, many investigators have conducted genome-wide sweeps and have used a variety of statistical criteria to judge whether identified regions of the human genome have a significant influence on the expression of some quantitative trait, such as measurements on patients with Alzheimer's disease. From the point of view of quantitative genetics, the regions of a genome that have some influence on a quantitative trait may be viewed as loci, and variations among these loci at the DNA level, such as nucleotide substitutions or other markers, may be used as working definitions of alleles, and, therefore, can be used to determine whether an individual carries a particular allele at some locus. Given such data, an investigator can identify the genotype of each individual in a study, with respect to the loci under consideration as well as the two alleles present at each locus in a diploid species such as man. This ability to use these working definitions to identify the genotype of each individual in a sample results in a significant change in the working paradigm of the subfield of quantitative genetics called variance and covariance analysis, because effects and components of variance and covariance my be estimated directly in a sense that will be described in detail in the paper.

*The content of this chapter was published in an online journal: Mode C. J. (2014) Estimating statistical measures of pleiotropic and epistatic effects in the genomic era. *International Journal of Statistics and Probability*, Vol. 3, No. 2. ISSN 1927-7032. The content of this chapter is a pdf image of the LaTeX version of the paper that was accepted for publication.

1. Introduction

Following the sequencing of the human genome in 2000 as well as that of other species, sequencing technology has advanced to the point at which it is becoming financially feasible — see Church (2006) [6] — to sequence the genomes of individuals in samples under study by an investigator or teams of investigators. This technological advance has led investigators to conduct genome-wide sweeps to search for regions of the human genome as well as those of other species such that there is evidence to suggest that these regions are implicated in the expression of complex quantitative traits. In an interesting paper by Stranger *et al.* (2010) [23], the impact of genome-wide association studies on the genetics of complex traits is discussed in depth. Among these complex traits are Alzheimer's disease (AD) and immune-mediated diseases such as rheumatoid arthritis. For the case of AD, in a recent paper Raj *et al.* (2012) [21] have reported that 11 regions of the human genome are involved in susceptibility to this disease, and, moreover, there is evidence that four of these regions form a protein–protein interaction network that is under natural selection. Similarly, in a paper by Rossin *et al.* (2011) [22], it has been found that proteins encoded in genomic regions associated with immune-mediated disease physically interact and this interaction may also suggest some biological mechanisms underlying such diseases. In some samples of patients with AD, the genome of each individual in the sample has been sequenced so that, in principle, the genotype of each individual may be determined with respect to 11 loci, regions of the human genome, or in combinations of these loci, whenever there are working definitions of at least two alleles at each locus.

This ability to identify the genotype of each individual in a sample with respect to some quantitative trait provides a concrete basis for extending some of the techniques of classical quantitative genetics into the era of sequenced genomes. In classical quantitative genetics, however, the loci under consideration as well as the alleles at each locus were unknown to an investigator so that these notions were treated abstractly and the parameters of a model could only be estimated indirectly. Briefly, a variety of analysis of variances and covariance procedures were used to estimate parameters of interest indirectly. A concrete example of such procedures may be found in the material accompanying Table 1 in Mode and Robinson (1959) [18]. But, as will be shown in subsequent sections, whenever the genotype of each individual in a sample may be identified with respect to some loci with at least two alleles at each locus, the parameters of the model may be estimated in a straightforward manner based on elementary methods of statistical estimation.

As is recognized by many who at some time during their careers have worked in the field of quantitative genetics, the subject known as components of variance analysis began with the publication of a paper on correlations among relatives on the supposition of Mendelian inheritance by R. A. Fisher (1918) [9]. In this paper, Fisher attempted to reconcile existing biometrical theories with Mendelian genetics, which led to describing genetic variation in terms of components of variance. Evidently at that time, the use of the word "variance" was relatively new, because at the beginning of the paper Fisher emphasized the word by representing it in upper case letters. Subsequently, the ideas of Fisher were extended to include epistasis and other interactions among alleles in the seminal papers of Cockerham (1954) [7] and Kempthorne (1954) [12]. The techniques introduced in these papers have been applied in the current genomic era. Examples of the ideas introduced by Cockerham have been used in the paper Kao *et al.* (2002) [11], and those of Kempthorne have been used and extended in the paper Mao *et al.* (2006) [17]. The ideas of Kempthorne were also used and extended in the paper of Mode and Robinson (1959) as well as in unpublished lecture notes by the author written and presented during the period 1960 to 1966. Furthermore, the roots of the ideas presented in this paper are extensions of some of the unpublished material in the lecture notes compiled by the author during the period 1960 to 1966. Many of the themes of statistical genetics as they existed during the 1950s were summarized and extended in Kempthorne's well-known book Kempthorne (1957) [13].

During the years following Fisher's seminal work and those cited above by Cockerham and Kempthorne, an extensive literature on quantitative genetics has evolved. It is beyond the scope of this paper to review this literature and in what follows a few books on the subject will be cited. A book that has been very popular with quantitative geneticists is that of Falconer and MacKay (1996) [8] as well as earlier editions. Another book of interest on quantitative genetics is that of Bulmer (1980) [5]. Both of these books contain extensive lists of references on quantitative genetics. A more recent book on genetics and analysis of quantitative traits is that of Lynch and Walsh (1998) [16]. This is an influential tome, consisting of over 900 pages and containing what seems to be the most extensive treatment of the subject of quantitative genetics published in the 20th century. The principal focus of this book is a biological and evolutionary point of view along with an extensive use of applied statistical methods. There is also an extensive list of papers on quantitative genetics that a reader who is interested in quantitative genetics may wish to peruse. The book by Liu (1998) [15] on statistical genetics focuses on statistical genetics along with linkage, mapping and quantitative trait linkage (QTL) analysis. Two recent books on statistical genetics are those of Laird and Lange (2011) [14] and Wu, Ma and Casella (2010) [24]. It is suggested that if the reader is interested in an overview of the material that is being taught in courses in quantitative and statistical genetics or simply an introduction to these subjects, the books cited above be consulted.

2. Pleiotropism and the Phenotypic, Genetic and Environmental Covariance Matrices for the Case of One Autosomal Locus

Pleiotropism is a term used by geneticists to describe cases in which several traits, discrete or quantitative, seem to be governed by alleles at a single autosomal locus or a locus on chromosomes that govern the sex of an individual. Because quantitative traits are important in agriculture and medicine, the focus of attention in this paper will be quantitative traits, whose genetics seems to be governed by alleles at a single autosomal locus. Let $\mathbb{A}$ denote the set of alleles at some autosomal locus and let the symbols x and y denote alleles in the set $\mathbb{A}$. In what follows, the number of alleles in this set will be assumed to be finite. The genotype of an individual will be denoted by the symbol (x, y), where x and y are alleles in $\mathbb{A}$ contributed by the maternal and paternal parents, respectively. Let $\mathbb{G}$ denote the set of genotypes under consideration so that for every genotype (x, y) it follows that $(x, y) \in \mathbb{G}$. To take into account pleiotropic effects for $k \geq 2$ quantitative traits, let $W_1, W_2, \ldots, W_k$ be k random variables characterizing the phenotypic variation in these k traits among individuals in a population or sample under consideration. It will be assumed that these random variables take values in the sets $\mathfrak{R}_1, \mathfrak{R}_2, \ldots, \mathfrak{R}_k$ of quantitative measurements expressed in terms of real numbers. As a first step in developing a succinct notation that will be used extensively in what follows, let W denote a $k \times 1$ column vector whose components are the random variables $W_1, W_2, \ldots, W_k$. In what follows, the symbol $\mathbb{R}_k$ will denote the set $k \geq 2$ dimensional set real numbers, which is the set of all possible values that may be realized by the vector W, and let the symbol T denote the transpose of a vector or matrix.

As the technology for sequencing genomes of individuals continues to develop, it will become financially feasible to sequence the genomes of all individuals in a sample. Furthermore, in some cases, it has been possible to find evidence that some region of a genome has been implicated in the phenotypic expression of the $k \geq 2$ quantitative traits under consideration. Moreover, it may also be possible to differentiate the alleles in this genomic region so that the genotype (x, y) of each individual in a population or sample may be identified in terms of the bases or other characteristics of the DNA in a genomic region by identifying each of the alleles x and y. It will be at the discretion of an investigator as to whether the maternal and paternal alleles are identified if such information is indeed available. Actually, in what follows, all that is necessary is that two alleles carried by an individual at a particular locus be identifiable. In any event, whenever it is possible to identify the genotype of each individual in a population or sample, it will be feasible to develop methods of statistical estimation that can take advantage of this information and provide more direct methods for estimating and drawing inferences about the parameters in a quantitative genetic model accommodating pleiotropic effects.

From the statistical point of view, the model under consideration will be a mixture of a discrete and a continuous multivariate distribution. Let

$$\mathbb{D}_{\text{Geno}} = \big(p(x, y) | (x, y \in G)\big) \tag{2.1}$$

denote the genotypic distribution of the population under consideration, where $p(x, y) \geq 0$ for all $(x, y) \in G$ and

$$\sum_{(x,y)} p(x, y) = 1. \tag{2.2}$$

By way of interpreting the distribution $\mathbb{D}_{\text{Geno}}$, let the random pair (X, Y) denote a genotype of an individual chosen at random from a population. Then,

$$P[(X, Y) = (x, y)] = p(x, y).$$

It should be mentioned that if a sample of individuals resulted from matings of parents whose genotypes were known, then, for the case of one locus, the theoretical genotypic distribution could be predicted by the well-known Mendelian theory. But, in general, in most samples of sequenced individuals, the genotypes of their parents are not known so that there would not be a theoretical basis for predicting the form of the genotypic distribution.

Given that $(X, Y) = (x, y)$, let $f(w|(x, y))$ denote the conditional density of the random vector W. It will be assumed that this distribution has the expectation vector

$$E[W|(x, y)] = \boldsymbol{\mu}(x, y) = \int_{\mathbb{R}_k} f(w|(x, y))dw \tag{2.3}$$

and the covariance matrix

$$E\left[(W - \boldsymbol{\mu}(x, y))(W - \boldsymbol{\mu}(x, y))^T|(x, y)\right] = \boldsymbol{\Psi}(x, y) \tag{2.4}$$

with finite elements for all genotypes $(x, y) \in \mathbb{G}$. From these definitions, it follows that the unconditional $k \times 1$ expectation vector of the population is

$$\boldsymbol{\mu} = \sum_{(x,y)} p(x, y)\boldsymbol{\mu}(x, y), \tag{2.5}$$

and the $k \times k$ unconditional covariance matrix of the population is

$$\boldsymbol{\Psi} = \sum_{(x,y)} p(x, y)\boldsymbol{\Psi}(x, y). \tag{2.6}$$

By definition, the marginal or gene frequency of the maternal allele x in the population is

$$p(x) = \sum_{y} p(x, y) \tag{2.7}$$

for all $x \in \mathbb{A}$. The marginal frequency $p(y)$ of the paternal allele y is defined similarly. A population is said to be in a Hardy–Weinberg equilibrium if

$$p(x, y) = p(x)p(y) \tag{2.8}$$

for all genotypes $(x, y) \in \mathbb{G}$.

In the formulation under consideration, vectors that are measures of first and second order effects of the alleles x and y on the k quantitative traits under consideration will be defined in terms of conditional expectations of the mean vector $\boldsymbol{\mu}(x, y)$ for each genotypic $(x, y) \in \mathbb{G}$ with respect to the genotypic distribution. In general, one would not expect that a population would be in a Hardy–Weinberg equilibrium so that for any particular sample an investigator may wish to test the hypothesis that a population was in this type of equilibrium. To accommodate the case in which a population is not in a Hardy–Weinberg equilibrium, vectors of first order effects will be defined in terms of conditional distributions. By definition, for $p(x) \neq 0$, the conditional distribution of the allele $y \in \mathbb{A}$, given x, is

$$p(y|x) = \frac{p(x, y)}{p(x)}. \tag{2.9}$$

Thus, the conditional expectation of the conditional mean vector $\boldsymbol{\mu}(x, y)$, given x, is

$$\boldsymbol{\mu}(x) = \sum_y p(y|x)\boldsymbol{\mu}(x, y). \tag{2.10}$$

It is interesting to note that if the population is in a Hardy–Weinberg equilibrium, then

$$\boldsymbol{\mu}(x) = \sum_y \frac{p(x)p(y)}{p(x)}\boldsymbol{\mu}(x, y) = \sum_y p(y)\boldsymbol{\mu}(x, y). \tag{2.11}$$

The conditional expectation of $\boldsymbol{\mu}(x, y)$, given y such that $p(y) \neq 0$, is defined similarly.

The first order effect of the allele x is defined by the vector equation

$$\boldsymbol{\alpha}(x) = \boldsymbol{\mu}(x) - \boldsymbol{\mu} \tag{2.12}$$

for all $x \in \mathbb{A}$. Similarly, the first order effect of the allele y is defined by

$$\boldsymbol{\alpha}(y) = \boldsymbol{\mu}(y) - \boldsymbol{\mu} \tag{2.13}$$

for all $y \in \mathbb{A}$. The second order effect of the alleles x and y due to their interaction is defined by the vector equation

$$\boldsymbol{\alpha}(x, y) = \boldsymbol{\mu}(x, y) - \boldsymbol{\mu} - \boldsymbol{\alpha}(x) - \boldsymbol{\alpha}(y) \tag{2.14}$$

for all genotypes $(x, y) \in \mathbb{G}$. Equivalently,

$$\boldsymbol{\mu}(x, y) = \boldsymbol{\mu} + \boldsymbol{\alpha}(x) + \boldsymbol{\alpha}(y) + \boldsymbol{\alpha}(x, y) \tag{2.15}$$

for all genotypes $(x, y) \in \mathbb{G}$. This equation reminds one of an analysis of variance model of a multivariate experimental design with two factors. If $\boldsymbol{\alpha}(x, y) = \mathbf{0}$, the zero vector, for all genotypes $(x, y) \in \mathbb{G}$, then the alleles x and y act additively with respect to the k quantitative measurements. But if $\boldsymbol{\alpha}(x, y) \neq \mathbf{0}$ for some genotype (x, y), then there is interaction among the alleles.

In a similar fashion, with respect to the random vector W with k phenotypic measurements for genotypes (x, y), it can be seen that the vector equation

$$W = \boldsymbol{\mu} + (\boldsymbol{\mu}(x, y) - \boldsymbol{\mu}) + (W - \boldsymbol{\mu}(x, y)) \tag{2.16}$$

is valid for all genotypes $(x, y) \in \mathbb{G}$. By definition, the total phenotypic covariance matrix with respect to k quantitative traits in the population is

$$\text{cov}_P[W] = \sum_{(x,y)} p(x, y) E\left[(W - \boldsymbol{\mu})(W - \boldsymbol{\mu})^T |(x, y)\right], \tag{2.17}$$

where the conditional expectation is taken with respect to the multivariate conditional density $f(w|(x, y))$ for each genotype (x, y). The covariance matrix

$$\text{cov}_G[W] = \sum_{(x,y)} p(x, y)(\boldsymbol{\mu}(x, y) - \boldsymbol{\mu})(\boldsymbol{\mu}(x, y) - \boldsymbol{\mu})^T, \tag{2.18}$$

measuring the covariation of the mean vectors $\boldsymbol{\mu}(x, y)$ for genotypes governing $k \geq 2$ traits around the mean vector $\boldsymbol{\mu}$ for the population, is called the genetic covariance matrix. Finally, the covariance matrix

$$\text{cov}_E[W] = \boldsymbol{\Psi} = \sum_{(x,y)} p(x, y) E\left[(W - \boldsymbol{\mu}(x, y))(W - \boldsymbol{\mu}(x, y))^T |(x, y)\right] \tag{2.19}$$

is called the environmental covariance matrix in the context of equation (2.16). Observe that it is the same matrix as that defined in equation (2.6).

From equation (2.16) it follows that

$$\begin{aligned}
&E\left[(W - \boldsymbol{\mu})(W - \boldsymbol{\mu})^T |(x, y)\right] \\
&\quad = E\left[(\boldsymbol{\mu}(x, y) - \boldsymbol{\mu})(\boldsymbol{\mu}(x, y) - \boldsymbol{\mu})^T |(x, y)\right] \\
&\qquad + E\left[(\boldsymbol{\mu}(x, y) - \boldsymbol{\mu})(W - \boldsymbol{\mu}(x, y))^T |(x, y)\right]
\end{aligned}$$

$$+E\left[(W-\boldsymbol{\mu}(x,y))(\boldsymbol{\mu}(x,y)-\boldsymbol{\mu})^T|(x,y)\right]$$
$$+E\left[(W-\boldsymbol{\mu}(x,y))(W-\boldsymbol{\mu}(x,y))^T|(x,y)\right]. \quad (2.20)$$

But

$$\begin{aligned} &E\left[(W-\boldsymbol{\mu}(x,y))(\boldsymbol{\mu}(x,y)-\boldsymbol{\mu})^T|(x,y)\right] \\ &\quad = E\left[(W-\boldsymbol{\mu}(x,y))|(x,y)\right](\boldsymbol{\mu}(x,y)-\boldsymbol{\mu})^T \\ &\quad = \mathbf{0}_{k\times 1}(\boldsymbol{\mu}(x,y)-\boldsymbol{\mu})^T = \mathbf{0}_{k\times k}, \end{aligned} \quad (2.21)$$

where, as indicated, $\mathbf{0}_{k\times 1}$ is a $k \times 1$ of zeros and $\mathbf{0}_{k\times k}$ is a $k \times k$ matrix of zeros. By way of validating this last step in (2.21), let X and Y be one-dimensional random variables, taking values in the set of real numbers. Then, it is well known that $E[XY|X] = XE[Y|X]$. Moreover, it can be shown that if X and Y are matrices such that the product XY is well defined, then the equation $E[XY|X] = XE[Y|X]$ also is valid for matrices. Furthermore, it can be shown that $E[XY|Y] = E[X|Y]Y$ is also valid for matrices. It was this well-known property that was used to justify the second expression in equation (2.21). These properties will also be used to show that the third term on the right in equation (2.20) is equal to $\mathbf{0}_{k\times k}$, a $k \times k$ matrix of zeros. If $f(X)$ is a matrix-valued function of a matrix X, then it can also be shown that $E\left[f(X)|X\right] = f(X)$.

Therefore, equation (2.20) reduces to

$$\begin{aligned} &E\left[(W(x,y)-\boldsymbol{\mu})(W(x,y)-\boldsymbol{\mu})^T|(x,y)\right] \\ &\quad = (\boldsymbol{\mu}(x,y)-\boldsymbol{\mu})(\boldsymbol{\mu}(x,y)-\boldsymbol{\mu})^T \\ &\qquad + E\left[(W-\boldsymbol{\mu}(x,y))(W-\boldsymbol{\mu}(x,y))^T|(x,y)\right]. \end{aligned} \quad (2.22)$$

By multiplying this equation by $p(x,y)$ and summing over all genotypes $(x,y) \in \mathbb{G}$, it follows that

$$\mathbf{cov}_P[W] = \mathbf{cov}_G[W] + \mathbf{cov}_E[W]. \quad (2.23)$$

It is interesting to note that this equation was derived by using properties of conditional expectations and is free as to any assumptions that may be made about the distributions of the terms on the right in equation (2.16). For example, in classical quantitative genetics, for either one trait or many with pleiotropic effects, it was often assumed that the genetic effects $(\boldsymbol{\mu}(x,y)-\boldsymbol{\mu})$ and environmental effects $(W-\boldsymbol{\mu}(x,y))$ were distributed independently. But, as can be seen from the derivation of (2.23) just described, such assumptions are not necessary. It is also important to note that, even though the equation was derived under the assumption that only one autosomal locus was under consideration, equation (2.23) would also be valid under the assumption that two or more autosomal loci were under consideration, but the formal details of a proof of this statement will not be given here.

As is well known, the principal diagonal elements of any covariance matrix are the variances of the elements of a random vector W under consideration. In particular, let $\mathrm{var}_P[W_\nu]$ denote the phenotypic variance for trait ν in the random vector W. Similarly, let $\mathrm{var}_G[W_\nu]$ and $\mathrm{var}_E[W_\nu]$ denote, respectively, the genetic and environmental variances of trait $\nu = 1, 2, \ldots, k$. Then, from (2.23), it follows that

$$\mathrm{var}_P[W_\nu] = \mathrm{var}_G[W_\nu] + \mathrm{var}_E[W_\nu] \tag{2.24}$$

for every trait ν. A measure of the heritability of a trait that has used extensively by many investigators is the ratio

$$H_\nu = \frac{\mathrm{var}_G[W_\nu]}{\mathrm{var}_G[W_\nu] + \mathrm{var}_E[W_\nu]} \tag{2.25}$$

for every trait $\nu = 1, 2, \ldots, k$. In an actual application involving the analysis of data based in the structure presented in this section, an investigator may wish to estimate the fraction H_ν for every trait $\nu = 1, 2, \ldots, k$. Even though many investigators have used a ratio of form (2.25) to estimate the heritability of a quantitative trait, the details in the calculations may vary among investigators. It is suggested that if the reader is interested in pursuing these details, the books cited in the introduction be consulted. Three recent papers containing applications of the concept of heritability are those of Yang *et al.* (2010) [25], Zaitlen *et al.* (2013) [26] and Price *et al.* (2011) [20]. It should also be mentioned that, even though only one autosomal locus has been under consideration is this section, equations (2.23) and (2.24) are also be valid if two or more autosomal loci were under consideration. Hence, the methods outlined in this section are also valid for any combination of two or more autosomal loci.

3. Partitioning the Genetic Covariance Matrix into Component Covariance Matrices for the Case of One Autosomal Locus

From equation (2.15), it can be seen from equation (2.16) that the $k \times 1$ vector expression $\mu(x, y) - \boldsymbol{\mu}$ may be expressed as the $k \times 1$ vector equation

$$\mu(x, y) - \boldsymbol{\mu} = \boldsymbol{\alpha}(x) + \boldsymbol{\alpha}(y) + \boldsymbol{\alpha}(x, y) \tag{3.1}$$

for every genotype $(x, y) \in \mathbb{G}$. From the definition of the vector $\boldsymbol{\alpha}(x)$ in equation (2.12), it can be seen that its expectation with respect to the genotypic distribution $\mathbb{D}_{\text{Geno}}$ is

$$E_{\mathbb{D}_{\text{Geno}}}[\boldsymbol{\alpha}(x)] = \sum_{(x,y)} p(x, y)\boldsymbol{\alpha}(x) = \sum_{x} p(x)\boldsymbol{\alpha}(x) = \mathbf{0}, \tag{3.2}$$

a $k \times 1$ zero vector. Similarly, it can be shown that

$$E_{\mathbb{D}_{\text{Geno}}}[\boldsymbol{\alpha}(y)] = \mathbf{0}, \tag{3.3}$$

$$E_{\mathbb{D}_{\text{Geno}}}[\boldsymbol{\alpha}(x, y)] = \mathbf{0}, \tag{3.4}$$

where $\mathbf{0}$ is a $k \times 1$ vector of zeros.

By definition, the additive genetic covariance matrix is

$$\text{cov}_A[W] = E_{\mathbb{D}_{\text{Geno}}}\left[(\boldsymbol{\alpha}(x)(\boldsymbol{\alpha}(x))^T) + (\boldsymbol{\alpha}(y)(\boldsymbol{\alpha}(y))^T)\right] \tag{3.5}$$

and the intra-allelic interaction covariance matrix is defined as

$$\text{cov}_{\text{IAI}}[W] = E_{\mathbb{D}_{\text{Geno}}}\left[(\boldsymbol{\alpha}(x, y)(\boldsymbol{\alpha}(x, y))^T)\right]. \tag{3.6}$$

Observe that the matrices in (3.4) and (3.5) are of order $k \times k$.

If a population is not in a Hardy–Weinberg equilibrium, then it is necessary to define cross-covariance matrices. For example, the covariance matrix of the vectors $\boldsymbol{\alpha}(x)$ and $\boldsymbol{\alpha}(y)$ is defined by the matrix expression

$$E_{\mathbb{D}_{\text{Geno}}}\left[(\boldsymbol{\alpha}(x))(\boldsymbol{\alpha}(y))^T\right]. \tag{3.7}$$

The covariance matrices of the pairs of vectors $\boldsymbol{\alpha}(x)$ and $\boldsymbol{\alpha}(x, y)$ as well as $\boldsymbol{\alpha}(y)$ and $\boldsymbol{\alpha}(x, y)$ are defined similarly. If the population is in a Hardy–Weinberg equilibrium, then it can be shown that all cross-covariance matrices are zero matrices, and it follows that the genetic covariance matrix may be partitioned into additive and intra-allelic interaction covariance matrices. Thus, when a population is in a Hardy–Weinberg equilibrium, the $k \times k$ matrix equation

$$\mathrm{cov}_G[W] = \mathrm{cov}_A[W] + \mathrm{cov}_{IAI}[W] \tag{3.8}$$

is valid.

If, however, the population is not in a Hardy–Weinberg equilibrium, then the analogue of equation (3.8) is more complicated. In this case, it will be helpful to express equation (3.8) in a more general form. To that end, consider the $3k \times 1$ column vector

$$\mathbf{\Phi}(x, y) = \begin{pmatrix} \boldsymbol{\alpha}(x) \\ \boldsymbol{\alpha}(y) \\ \boldsymbol{\alpha}(x, y) \end{pmatrix} \tag{3.9}$$

and let

$$\mathbf{\Psi}(x, y) = \mathbf{\Phi}(x, y)\mathbf{\Phi}^T(x, y) \tag{3.10}$$

denote a $3k \times 3k$ matrix of $k \times k$ submatrices on the principal diagonal and $k \times k$ cross-product matrices off the principal diagonal. The expectation $\mathbf{\Psi}_G$ of this matrix with respect to the genotypic distribution is defined by

$$\mathbf{\Psi}_G = \sum_{(x,y)} p(x, y)\mathbf{\Psi}(x, y). \tag{3.11}$$

In general, when the population is not in a Hardy–Weinberg equilibrium, it will be helpful to represent the matrix $\mathbf{\Psi}_G$ in the succinct partitioned form

$$\mathbf{\Psi}_G = \begin{pmatrix} \mathbf{\Psi}_G(1,1) & \mathbf{\Psi}_G(1,2) & \mathbf{\Psi}_G(1,3) \\ \mathbf{\Psi}_G(2,1) & \mathbf{\Psi}_G(2,2) & \mathbf{\Psi}_G(2,3) \\ \mathbf{\Psi}_G(3,1) & \mathbf{\Psi}_G(3,2) & \mathbf{\Psi}_G(3,3) \end{pmatrix}, \tag{3.12}$$

where every submatrix is of order $k \times k$. In terms of this matrix, it follows from the vector in (3.9) that the additive covariance matrix $\mathrm{cov}_A[W]$ in (3.8) is

$$\mathrm{cov}_A[W] = \mathbf{\Psi}_G(1, 1) + \mathbf{\Psi}_G(2, 2), \tag{3.13}$$

and the intra-allelic interaction covariance matrix in (3.9) is

$$\mathrm{cov}_{IAI}[W] = \mathbf{\Psi}_G(3, 3). \tag{3.14}$$

Let $k \times k$ matrix $R_G[W]$ denote the sum

$$R_G[W] = \sum_{i \neq j} \Psi_G(i,j). \tag{3.15}$$

Given these definitions, the desired extension of equation (3.9) to the case of a population is not in a Hardy–Weinberg equilibrium has the form

$$\mathbf{cov}_G[W] = \mathbf{cov}_A[W] + \mathbf{cov}_{\text{IAI}}[W] + R_G[W]. \tag{3.16}$$

Let X denote any $k \times 1$ random vector with a covariance matrix $\mathbf{cov}[X] = (\text{cov}_{ij}[X])$ such that all its elements are finite. If $i = j$, then $\text{cov}_{ii}[X] = \text{var}[X_i]$, where X_i is the ith component of the vector X. In terms of this notation, the νth component on the principal diagonal is matrix equation (3.16) has the form

$$\text{var}_G[W_\nu] = \text{var}_A[W_\nu] + \text{var}_{\text{IAI}}[W_\nu] + \text{cov}_G[W_\nu], \tag{3.17}$$

for $\nu = 1, 2, \ldots, k$, where $\text{cov}_G[W_\nu]$ is the element of the matrix $R_G[W]$ in the row corresponding to position $\nu\nu$. For each trait $\nu = 1, 2, \ldots, k$, an investigator may wish to compute that ratio

$$r_A(\nu) = \frac{\text{var}_A[W_\nu]}{\text{var}_G[W_\nu]} \tag{3.18}$$

as a measure of the contribution of the additive genetic variance $\text{var}_A[W_\nu]$ to the total genetic $\text{var}_G[W_\nu]$. Moreover, it is interesting to note that the ratio

$$r_{\text{IAI}}(\nu) = \frac{\text{var}_{\text{IAI}}[W_\nu]}{\text{var}_G[W_\nu]} \tag{3.19}$$

and may be interpreted as a measure of the contribution of the intra-allelic interaction variance to the total genetic variance for trait ν. An investigator may also be interested in computing the ratio

$$1 - r_A(\nu) - r_{\text{IAI}}(\nu) = \frac{\text{cov}_G[W_\nu]}{\text{var}_G[W_\nu]}$$

as a measure of the departure from a Hardy–Weinberg equilibrium. Observe that if $\text{cov}_G[W_\nu]$ is close to zero, then a Hardy–Weinberg disequilibrium would have a minimal effect on estimating $r_A(\nu)$ or $r_{\text{IAI}}(\nu)$. But, if $|\text{cov}_G[W_\nu]|$ is relatively large, then the effect of a disequilibrium could be significant in estimating either of these ratios.

Correlation matrices may also be of interest to some investigators. For example, for the case of the genetic covariance matrix $\mathbf{cov}_G[W]$ in (3.16), an investigator may wish to look at the correlation matrix derived from this covariance matrix. Let $\sigma_G^2(\nu) = \text{var}_G[W_\nu]$ denote the variance of trait ν and let $\sigma_G(\nu, \nu') = \text{cov}_G[W_\nu, W_{\nu'}]$ denote the covariance of traits ν and ν' such that $\nu \neq \nu'$. Then, the genetic correlation coefficient of traits ν and ν' has the form

$$\rho_G(\nu, \nu') = \frac{\text{cov}_G\,[W_\nu, W_{\nu'}]}{\sigma_G(\nu)\sigma_G(\nu')}. \tag{3.20}$$

If $\nu = \nu'$,

$$\rho_G(\nu, \nu) = \frac{\sigma_G^2(\nu)}{\sigma_G^2(\nu)} = 1. \tag{3.21}$$

Then, the $k \times k$ genetic correlation has the form

$$\textbf{corr}_G[W] = \left(\rho_G(\nu, \nu')\right). \tag{3.22}$$

Observe that all the components of the principal diagonal of this matrix are 1. If an investigator were also interested in the correlation matrices corresponding to the covariance matrices $\textbf{cov}_A[W]$ and $\textbf{cov}_{IAI}[W]$ in (3.16), then the correlation matrices for these covariance matrices could also be computed. For all the matrices just mentioned, as well as those that may arise in subsequent sections of this paper, the value of correlation coefficients may be interpreted as a measure of pleiotropic effects.

4. Estimation of Mean Genetic Vector and Covariance Matrices from Data for the Case of One Locus

It should be stated at the outset that one could assume that the multivariate $k \times 1$ random vectors $W(x, y)$ under consideration for each genotype were distributed independently with multivariate normal distributions with expectation vector $\boldsymbol{\mu}(x, y)$ and covariance matrix $\boldsymbol{\Psi}(x, y)$ for all genotypes $(x, y) \in \mathbb{G}$. But, even though such assumptions may be valuable in coming to grips with some of the problems that will be encountered in subsequent sections, particularly those involving multiple comparisons, a decision was made to minimize the formalism presented in this paper with the hope that a less involved notation would attract more readers with interests in statistical genetics and related disciplines. Thus, for the most part, the methods of estimation to be described in this section are extensions of the method of moments, which is among the most primitive methods used in statistical estimation, but is, nevertheless, still effective in many situations in which problems of estimation of parameters arise. If, however, the reader is interested in pursuing theoretical treatments of multivariate statistics, it is suggested that the books by Anderson (1984) [1] and Muirhead (1982) [19] be consulted.

For each genotype $(x, y) \in \mathbb{G}$, suppose that there are $n(x, y) \geq 2$ individuals of genotype (x, y) that are measured with respect to the k quantitative traits under consideration, and let the random $k \times 1$ vectors $\boldsymbol{W}_\nu(x, y)$ for $\nu = 1, 2, \ldots, n(x, y)$ denote the observed data for the $n(x, y)$ individuals of genotype (x, y). It will be assumed that these random vectors are distributed independently and that each vector has the same distribution as the phenotypic vector $\boldsymbol{W}(x, y)$ for individuals of genotype (x, y) defined in Section 2. Then, the random vector

$$\widehat{\boldsymbol{\mu}}(x, y) = \frac{1}{n(x, y)} \sum_{\nu=1}^{n(x,y)} \boldsymbol{W}_\nu(x, y) \tag{4.1}$$

is an estimator of the $k \times 1$ expectation vector $\boldsymbol{\mu}(x, y)$ defined in (2.3) for all genotypes $(x, y) \in \mathbb{G}$. Given the estimator $\widehat{\boldsymbol{\mu}}(x, y)$, it can be shown that the random matrix

$$\widehat{\boldsymbol{\Psi}}(x, y) = \frac{1}{n(x, y) - 1} \sum_{\nu=1}^{n(x,y)} (\boldsymbol{W}_\nu(x, y) - \widehat{\boldsymbol{\mu}}(x, y))(\boldsymbol{W}_\nu(x, y) - \widehat{\boldsymbol{\mu}}(x, y))^T \tag{4.2}$$

is an estimator of the covariance matrix $\boldsymbol{\Psi}(x, y)$ defined in (2.4) for all genotypes $(x, y) \in \mathbb{G}$. It also is well known that both these estimators are unbiased in the sense that $E[\widehat{\boldsymbol{\mu}}(x, y)|(x, y)] = \boldsymbol{\mu}(x, y)$ and $E[\widehat{\boldsymbol{\Psi}}(x, y)|(x, y)] = \boldsymbol{\Psi}(x, y)$ for all genotypes $(x, y) \in \mathbb{G}$.

Let

$$n = \sum_{(x,y)} n(x, y) \tag{4.3}$$

denote that total number of observed individuals. Then,

$$\widehat{p}(x, y) = \frac{n(x, y)}{n} \tag{4.4}$$

is an estimator of the frequency of genotype (x, y) in the population. If it is assumed that the observations $\{n(x, y)|(x, y) \in \mathbb{G})\}$ are a sample from a multinomial distribution with the genotypic distribution $\mathbb{D}_{\text{Geno}}$ in (2.1) as its probabilities with sample size n, then

$$E[\widehat{p}(x,y)] = \frac{1}{n}E[n(x,y)] = \frac{1}{n}np(x,y) = p(x,y) \tag{4.5}$$

for all genotypes $(x,y) \in \mathbb{G}$ so that $\widehat{p}(x,y)$ is an unbiased estimator of $p(x,y)$ under these assumptions. From (2.5) it follows that the random variable

$$\widehat{\boldsymbol{\mu}} = \sum_{(x,y)} \widehat{p}(x,y)\widehat{\boldsymbol{\mu}}(x,y) \tag{4.6}$$

is an estimator of the unconditional expectation vector $\boldsymbol{\mu}$ defined in (2.5). Similarly, the random matrix

$$\widehat{\boldsymbol{\Psi}} = \sum_{(x,y)} \widehat{p}(x,y)\widehat{\boldsymbol{\Psi}}(x,y) \tag{4.7}$$

is an estimator of the unconditional covariance matrix $\boldsymbol{\Psi}$ defined in (2.6).

Given the estimators described above, it follows that the covariance matrices on the right side of equation (2.23) can be estimated. For example, from the definition of the environmental matrix in equations (2.22) and (2.23), it follows that

$$\widehat{\mathbf{cov}}_E[W] = \widehat{\boldsymbol{\Psi}}_E = \widehat{\boldsymbol{\Psi}}; \tag{4.8}$$

see (4.7). Similarly, the random matrix

$$\widehat{\mathbf{cov}}_G[W] = \sum_{(x,y)} \widehat{p}(x,y)(\widehat{\boldsymbol{\mu}}(x,y) - \widehat{\boldsymbol{\mu}})(\widehat{\boldsymbol{\mu}}(x,y) - \widehat{\boldsymbol{\mu}})^T \tag{4.9}$$

is an estimator of the genetic matrix in (2.23). From these results, it follows from equation (2.23) that the random matrix

$$\widehat{\mathbf{cov}}_P[W] = \widehat{\mathbf{cov}}_G[W] + \widehat{\mathbf{cov}}_E[W] \tag{4.10}$$

is an estimator of the phenotypic covariance matrix in (2.23). Given the estimates of covariance matrices on the right in (4.10), an investigator may wish to estimate the measure of heritability of each trait, using the formula in equation (2.25).

Although the formal details will not be given here, it is easy to see that the effects defined in Section 2 as well as the variance components defined in Section 3 could be estimated directly without recourse to any analysis of variance procedure. This estimation procedure could be carried out under the assumption that the population was either in a Hardy–Weinberg equilibrium or not in such an equilibrium. In either case, all the covariance matrices defined in Section 3 could be estimated.

It is easy to see from (4.9) that the genetic covariance is symmetric, because

$$\left((\widehat{\mu}(x,y)-\widehat{\mu})(\widehat{\mu}(x,y)-\widehat{\mu})^T\right)^T = \left(\widehat{\mu}(x,y)-\widehat{\mu}\right)\left(\widehat{\mu}(x,y)-\widehat{\mu}\right)^T \quad (4.11)$$

for every genotype $(x,y) \in \mathbb{G}$. Therefore, since the transpose of a sum of matrices is the sum of the transpose of each matrix in (4.9) in the sum, it follows that

$$(\widehat{\mathbf{cov}}_G[W])^T = \widehat{\mathbf{cov}}_G[W] \quad (4.12)$$

so that $\widehat{\mathbf{cov}}_G[W]$ is a symmetric matrix and will thus have real eigenvalues. As suggested in Section 3, suppose that an investigator wished to estimate the genetic correlation matrix $\widehat{\mathbf{corr}}_G[W]$ defined in (3.22). As will be shown subsequently, if an estimate of the genetic covariance matrix is positive definite, then the corresponding estimates of the genetic correlation matrix will be such that all correlation coefficients $\widehat{\rho}$ will have values in the open interval $(-1,1)$.

Recall that for $k \geq 2$, a $k \times k$ symmetric real matrix A is positive definite if and only if $w^T A w > 0$ for all $k \times 1$ vectors $w \in \mathbb{R}_k$ such that $w \neq \mathbf{0}$, where $\mathbb{R}_k$ is the set of $k \times 1$ vectors of real numbers and $\mathbf{0}$ is the $k \times 1$ vector of zeroes. If for some vector $w \in \mathbb{R}_k$, such that $w \neq \mathbf{0}$ and $w^T A w = 0$, then the matrix A is said to be positive semidefinite. From this definition, it also follows that any square submatrix B_ν of A consisting of ν rows and ν columns is also positive definite for $\nu = 1, 2, \ldots, k-1$.

A positive definite matrix may be characterized in a number of ways. For example, it is well known that a matrix A is positive definite if and only its eigenvalues are all positive. If the reader is interested in finding a proof of this statement as well as a treatment of positive semidefinite matrices, it is suggested that the key phrase "positive definite matrices" be typed into an Internet search engine, where a wealth of material on linear algebra may be found in brief but reliable accounts. Among the many examples on the Internet is a Wikipedia article with this title. It is interesting to observe that it follows from its construction that the covariance matrix $\widehat{\mathbf{corr}}_G[W]$ is positive semidefinite. For example, for any $w \in \mathbb{R}_k$ it can be seen from (4.11) that

$w^T(\widehat{\mu}(x,y) - \widehat{\mu}) = u(x,y) = (\widehat{\mu}(x,y) - \widehat{\mu})^T w$, where $u(x,y) \in \mathbb{R}_1$ for all genotypes $(x,y) \in \mathbb{G}$. Therefore,

$$
\begin{aligned}
w^T \widehat{\mathrm{cov}}_G[W] w &= \sum_{(x,y)} \widehat{p}(x,y) w^T(\widehat{\mu}(x,y) - \widehat{\mu})(\widehat{\mu}(x,y) - \widehat{\mu})^T w \\
&= \sum_{(x,y)} \widehat{p}(x,y) u^2(x,y) \geq 0 \qquad (4.13)
\end{aligned}
$$

for all $w \in \mathbb{R}_k$.

At this juncture, it seems fitting to point out that a necessary property of any covariance matrix A is that it be positive semidefinite. To see that this statement is indeed true, let W be a $k \times 1$ column vector of random variables with values in $\mathbb{R}_k$ with covariance matrix A and let $a \in \mathbb{R}_k$. Then, let $Z = a^T W$ denote a random variable taking values in $\mathbb{R}_1$. Then, from the definition of the variance of a random variable, it follows that $\mathrm{var}[Z] \geq 0$ for all $a \in \mathbb{R}_k$. But it is well known that $\mathrm{var}[Z] = a^T A a \geq 0$ so that the matrix A must be positive semidefinite. Because the estimate of the genetic covariance matrix will always be positive semidefinite when the estimation procedure outlined in this section is used, it has a definite advantage when compared with that used by Mode and Robinson (1959) [18], which was carried out within a framework of an analysis of covariance table, because when using such procedures one could not, in general, guarantee that an estimated genetic covariance matrix was indeed always positive semidefinite and so on some occasions an investigator might get negative estimates of a variance component. If the reader is interested in the method of estimation just mentioned, it is suggested that the formal procedures accompanying Table 1 of the paper by Mode and Robinson just cited be consulted. It should also be mentioned that at the time this work was done in the 1950s, the presumed set of loci as well as the alleles at each locus governing the quantitative genetics of the traits under consideration were treated abstractly and without any knowledge of their locations in the genome of a species under study in an experiment. Consequently, the estimation procedure suggested in this section was not even conceivable the during the 1950s.

But, as mentioned above, it would be desirable to know whether the estimated genetic covariance matrix is indeed positive definite so that all estimated correlation coefficients would lie in the open interval $(-1, 1)$. It is recommended, therefore, that if an investigator is interested in estimating the genetic correlation matrix and using it to draw genetic inferences about a population, a first step would be that of computing the eigenvalues of the estimated genetic covariance matrix to determine whether all its eigenvalues are positive. Many software packages contain programs for calculating the eigenvalues of a square symmetric matrix so that a software developer would have little difficulty in finding a program to compute the eigenvalues of a covariance matrix. It should also be mentioned that if one or more of these eigenvalues are near zero, then the genetic covariance matrix would be nearly singular.

Among the several ways of characterizing a real positive definite matrix is a procedure known as Sylvester's criterion. For any symmetric $k \times k$ matrix $A = (a_{ij})$, define k determinants as follows:

$$D_1 = a_{11}, D_2 = \det \begin{bmatrix} a_{11} & a_{12} \\ a_{21} & a_{22} \end{bmatrix}, \ldots, D_k = \det A. \tag{4.14}$$

Then, the matrix A is positive definite if and only if $D_\nu > 0$ for all $\nu = 1, 2, \ldots, k$. From the computational point of view, Sylvester's criterion could also be used to check whether a square matrix is positive definite, but to carry out such a procedure a software developer would need to have access to a program based on vary efficient algorithms for finding the numerical value of the determinants in (4.13). As will be shown below, this criterion also has interesting implications for testing whether all estimated correlation coefficients $\widehat{\rho}$ belong to the open interval $(-1, 1)$.

For suppose that A is the estimate of the genetic covariance $\widehat{\mathbf{cov}}_G[W]$ matrix in (4.12) and suppose that it has been determined that the matrix is positive definite. To simplify the notation, represent the element of this matrix by $\widehat{\text{cov}}_G[W] = (\sigma_{ij})$. To further simplify the notation, the symbol $\widehat{\circ}$, indicating estimates of parameters are under consideration, has been omitted for the elements of the estimated genetic covariance matrix. In this matrix $\sigma_{ii} = \sigma_i^2$, the estimated variance of trait i for $i = 1, 2, \ldots, k$, and the covariance $\sigma_{ij} = \rho_{ij}\sigma_i\rho_j$, where $i \neq j$, ρ_{ij} is an estimate of the genetic correlation between traits i and j, and σ_i and σ_j are, respectively, the estimated genetic standard deviations for traits i and j. Because of symmetry, $\rho_{ij} = \rho_{ji}$ for all $i \neq j$. Because the estimated genetic covariance has, by assumption, been shown to be positive definite, it follows that Sylvester's criterion will be satisfied. In this notation, the first two determinates in Sylvester's criterion are $D_1 = \sigma_1^2 > 0$ and

$$D_2 = \det \begin{bmatrix} \sigma_1^2 & \rho_{12}\sigma_1\sigma_2 \\ \rho_{21}\sigma_2\sigma_1 & \sigma_2^2 \end{bmatrix}$$
$$= \sigma_1^2\sigma_2^2 - \rho_{12}^2\sigma_1^2\sigma_2^2 = \sigma_1^2\sigma_2^2\left(1 - \rho_{12}^2\right) > 0. \quad (4.15)$$

But, because $\sigma_1^2\sigma_2^2 > 0$, this equation implies that $1 - \rho_{12}^2 > 0$. Thus, $\rho_{12}^2 < 1$, which implies that $-1 < \rho_{12} < 1$. By proceeding systematically in this way by choosing 2×2 submatrices corresponding to two rows and two columns, it can be shown that all the estimated correlation coefficients $\widehat{\rho}_{ij}$ for all $i \neq j$ satisfy the condition $-1 < \widehat{\rho}_{ij} < 1$. For example, the 2×2 submatrix corresponding to rows and columns 1 and 3 of a $k \times k$ positive definite matrix has the form

$$\begin{bmatrix} \sigma_1^2 & \rho_{13}\sigma_1\sigma_3 \\ \rho_{31}\sigma_3\sigma_1 & \sigma_3^2 \end{bmatrix}, \quad (4.16)$$

which has the same form as the matrix in (4.14), which implies that

$$-1 < \widehat{\rho}_{13} < 1. \quad (4.17)$$

5. Measures of Pleiotropism and Epistasis for the Case of Two Autosomal Loci

Let $\mathbb{A}_1$ denote the set of alleles at locus 1 and let $\mathbb{A}_2$ denote the set of alleles at locus 2. It will be assumed that each of these sets contains at least two alleles but the total number of alleles at each locus is finite. Any genotype will be denoted by a symbol of the form (x_1, y_1, x_2, y_2), where the subscript, 1 or 2, denotes the locus, and the symbols x and y denote alleles contributed by the maternal and the paternal parent, respectively. In order to simplify the notation, a genotype will often be denoted by the single-letter symbol $z = (x_1, y_1, x_2, y_2)$. Let $\mathbb{G}$ denote the set of genotypes under consideration. The number of genotypes in the set $\mathbb{G}$ may be quite large, particularly if there are several alleles at each locus. Because the set of genotypes contains all possible combinations of alleles at each locus, the set $\mathbb{G}$ may be represented as the product set

$$\mathbb{G} = \mathbb{A}_1 \times \mathbb{A}_1 \times \mathbb{A}_2 \times \mathbb{A}_2. \quad (5.1)$$

For every genotype $z \in \mathbb{G}$, let $p(z)$ denote the frequency of genotype z in the population. Then, $p(z) \geq 0$ for all $z \in \mathbb{G}$ and

$$\sum_{z \in \mathbb{G}} p(z) = 1. \tag{5.2}$$

Just as in the case of one locus, the genotypic distribution will be denoted by

$$\mathbb{D}_{\text{Geno}} = \{p(z) | z \in \mathbb{G}\}. \tag{5.3}$$

As in the foregoing sections, to accommodate pleiotropism it will also be supposed that $k \geq 2$ traits are under consideration and that the random $k \times 1$ phenotypic vector W along with its distribution characterizes the observed variation with respect to k traits in a population or sample that is under consideration. Just as in Section 2, the $k \times 1$ conditional expectation or genetic vector

$$\boldsymbol{\mu}(z) = E[W | z] \tag{5.4}$$

along with the conditional covariance matrix

$$\boldsymbol{\Psi}(z) = E\left[(W - \boldsymbol{\mu}(z))(W - \boldsymbol{\mu}(z))^T | z\right], \tag{5.5}$$

which are defined for all genotypes $z \in \mathbb{G}$, will play essential roles in this section.

In Section 3, where only one autosomal locus was under consideration, an effect was introduced that was a measure of interactions of alleles at one locus. When two or more autosomal loci are under consideration, however, interactions among alleles at different loci will need to be accommodated in a model that will form the basis for a statistical analysis of data gathered in an experiment devoted to quantitative genetics. In classical genetics, interactions among alleles at different loci are referred to as epistasis. Briefly, the primary focus of attention in this section is to extend the results in Section 3 by defining effects, which are functions of the genetic expectations in (5.4), that are measures of epistasis as well as other interactions among sets of alleles. Just as in the foregoing sections, the unconditional genetic expectation vector

$$\boldsymbol{\mu} = \sum_{z \in \mathbb{G}} p(z) \boldsymbol{\mu}(z) \tag{5.6}$$

and unconditional covariance matrix

$$\boldsymbol{\Psi} = \sum_{z \in \mathbb{G}} p(z)\boldsymbol{\Psi}(z) \tag{5.7}$$

will play roles in what follows.

As a first step in defining effects for the case of two autosomal loci, it will be necessary to represent the genotypic distribution in a more explicit form. For every genotype $z = (x_1, y_1, x_2, y_2) \in \mathbb{G}$, let $p(x_1, y_1, x_2, y_2)$ denote its frequency in the population. Then, by definition,

$$p(x_1) = \sum_{(y_1, x_2, y_2)} p(x_1, y_1, x_2, y_2) \tag{5.8}$$

is the marginal frequency distribution of alleles $x_1 \in \mathbb{A}_1$ in the population. The marginal frequencies of alleles y_1, x_2 and y_2 are defined similarly. To lighten the notation, no subscripts, such as $p_1(x_1)$, will be attached to marginal distributions, because the subscript on each allele will denote the locus under consideration. Let $p(y_1), p(x_2)$ and $p(y_2)$ denote, respectively, the marginal frequencies of alleles y_1, x_2 and y_2. For the case of two autosomal loci, a population is said to be in linkage equilibrium if

$$p(x_1, y_1, x_2, y_2) = p(x_1)p(y_1)p(x_2)p(y_2) \tag{5.9}$$

for all genotypes $z = (x_1, y_1, x_2, y_2) \in \mathbb{G}$.

One would not, in general, expect that a population would be in linkage equilibrium for the case under consideration so that it becomes necessary to define $k \times 1$ vectors of effects using conditional distributions. For example, the conditional distribution of the alleles (y_1, x_2, y_2), given x_1, is

$$p(y_1, x_2, y_2 | x_1) = \frac{p(x_1, y_1, x_2, y_2)}{p(x_1)} \tag{5.10}$$

for $p(x_1) \neq 0$. It is easy to see that if (5.9) is satisfied, it follows that

$$p(y_1, x_2, y_2 | x_1) = p(y_1)p(x_2)p(y_2) \tag{5.11}$$

for all triples of alleles $(y_1, x_2, y_2) \in \mathbb{A}_1 \times \mathbb{A}_2 \times \mathbb{A}_2$. Let the $k \times 1$ vector $\boldsymbol{\mu}(x_1)$ denote the conditional expectation

$$\boldsymbol{\mu}(x_1) = \sum_{(y_1, x_2, y_2)} \boldsymbol{\mu}(x_1, y_1, x_2, y_2)p(y_1, x_2, y_2 | x_1) \tag{5.12}$$

for all $x_1 \in \mathbb{A}_1$. Then, just as in the one-autosomal-locus case, the vector $\boldsymbol{\alpha}(x_1)$ of effects for allele x_1 in the population is defined by

$$\boldsymbol{\alpha}(x_1) = \boldsymbol{\mu}(x_1) - \boldsymbol{\mu} \tag{5.13}$$

for all $x_1 \in \mathbb{A}_1$. Observe that

$$E_{\mathbb{D}_{\text{Geno}}}[\boldsymbol{\alpha}(x_1)] = \sum_{x_1} p(x_1)\boldsymbol{\alpha}(x_1) = \mathbf{0}, \tag{5.14}$$

a $k \times 1$ vector of zeroes. The first order effects just defined can easily be extended to define the $k \times 1$ vectors $\boldsymbol{\alpha}(y_1)$, $\boldsymbol{\alpha}(x_2)$ and $\boldsymbol{\alpha}(y_2)$, for all alleles $y_1 \in \mathbb{A}_1$ and all pairs $(x_2, y_2) \in \mathbb{A}_2 \times \mathbb{A}_2$.

For the case of two autosomal loci, it is possible to define many more effects than for the case of one autosomal locus. To provide a framework for defining and classifying these effects, it will be helpful to consider a set $\mathfrak{S} = (1, 2, 3, 4)$ of four positions that are occupied by alleles at the two loci under consideration. For example, for any genotype, positions 1 and 2 are occupied by alleles at locus 1, and positions 3 and 4 are occupied by alleles at locus 2. To get a grasp of how many effects that can be defined for the case under consideration, it is helpful to think of the class $\mathfrak{T}$ of all subsets of the set $\mathfrak{S}$. Among the sets in $\mathfrak{T}$ is φ, the empty set. To provide a means for describing effects that are measures of interactions of alleles at one locus or effects that are measures of epistatic interaction among alleles at two or more loci, it is useful to enumerate classes of subsets in the class $\mathfrak{T}$. For $\nu = 0, 1, 2, 3, 4$, let $\mathfrak{T}_\nu$ denote the class of subsets containing ν positions.

Thus, the class $\mathfrak{T}_0 = \{\varphi\}$ contains only the empty set, but the class $\mathfrak{T}_1$ consists of the singletons

$$\mathfrak{T}_1 = \{(1), (2), (3), (4)\}, \tag{5.15}$$

which were used to define the first order effects listed above. From elementary combinatorics, it is easy to see that the number of sets in the class $\mathfrak{T}_2$ is

$$\binom{4}{2} = 6.$$

In particular, the subsets in the class $\mathfrak{T}_2$ are

$$\mathfrak{T}_2 = \{(1, 2), (1, 3), (1, 4), (2, 3), (2, 4), (3, 4)\}. \tag{5.16}$$

Observe that the subclass of sets of positions

$$\mathfrak{T}_{2IAI} = \{(1,2),(3,4)\} \tag{5.17}$$

will form a basis for defining second order effects that are measures of intra-allelic interactions at loci 1 and 2, respectively. But the subclass of sets

$$\mathfrak{T}_{2EPI} = \{(1,3),(1,4),(2,3),(2,4)\} \tag{5.18}$$

provides a basis for defining second order epistatic effects among the two loci under consideration. There are

$$\binom{4}{3} = 4$$

subsets in the class $\mathfrak{T}_3$, which consists of the subsets

$$\mathfrak{T}_3 = \{(1,2,3),(1,2,4),(1,3,4),(2,3,4)\}. \tag{5.19}$$

As will be shown subsequently, the third order effects corresponding to the set in this class provide a basis for defining various types of effects that are measures of epistatic interactions. Finally, the class of sets

$$\mathfrak{T}_4 = \{\mathfrak{S} = (1,2,3,4)\} \tag{5.20}$$

contains all positions and provides a basis for defining fourth order effects that are measures of epistatic and intralocus interactions among any set of four alleles at the two loci under consideration.

From the foregoing discussion, it can be seen that a total of 15 effects may be defined for the case of two autosomal loci, and to derive a formula for computing each effect it would be necessary to consider 15 conditional probability distributions. Rather than deriving a formula for each of these 15 subsets of $\mathfrak{S}$, it will be helpful to set down general formulas for marginal and conditional distributions. Let

$$\mathfrak{E} = \bigcup_{\nu=1}^{3} \mathfrak{T}_\nu \tag{5.21}$$

denote the union of the classes of subsets of $\mathfrak{S}$ corresponding to effects under consideration. For every set $A \in \mathfrak{E}$, let A^c denote its complement with respect to $\mathfrak{S}$, and let $z(A)$ and $z(A^c)$ denote the sets of alleles in genotype z corresponding, respectively, to the positions in the sets A and A^c. For any genotype $z \in \mathbb{G}$, $z = z((A), z(A^c))$, where, by definition, the positions in the set A are fixed in the operations that follow. For example, the formula denotes

$$p(z(A)) = \sum_{z(A^c)} p(z(A), p(A^c)), \tag{5.22}$$

the marginal distribution of the alleles in the position of the set A, where the sum runs over all alleles in the positions of the set A^c. Thus, in this succinct notation,

$$p(z(A^c)|z(A)) = \frac{p(z(A), p(A^c))}{p(z(A))} \tag{5.23}$$

is the conditional distribution of the alleles in the positions in the set A^c, given the fixed alleles in the positions of the set A such that $p(z(A)) \neq 0$. Let $\boldsymbol{\mu}(z(A))$ denote the conditional expectation of the vector $\boldsymbol{\mu}(z)$, given the fixed alleles in the positions corresponding to the set A. Then, for every $A \in \mathfrak{E}$ let

$$\boldsymbol{\mu}(z(A)) = \sum_{z(A^c)} p(z(A^c)|z(A))\boldsymbol{\mu}(z((A), z(A^c))), \tag{5.24}$$

where the sum runs over all the alleles in the positions of the set A^c.

To illustrate using formula (5.24), consider those sets $A \in \mathfrak{T}_1$. If $A = \{1\}$, then $\boldsymbol{\mu}(z(A)) = \boldsymbol{\mu}(x_1)$ so that $\boldsymbol{\alpha}(x_1) = \boldsymbol{\mu}(x_1) - \boldsymbol{\mu}$ could be estimated for all alleles $x_1 \in \mathbb{A}_1$. The remaining first order effects, $\boldsymbol{\alpha}(y_1), \boldsymbol{\alpha}(x_2)$ and $\boldsymbol{\alpha}(y_2)$, could also be derived and estimated in a similar way, using formula (5.24). The number of sets in the class $\mathfrak{T}_2$ is six, so that there are six conditional expectations of the form $\boldsymbol{\mu}(A)$ for $A \in \mathfrak{T}_2$. Suppose, for example, that $A = \{x_1, y_1\}$. Then $\boldsymbol{\mu}(A) = \boldsymbol{\mu}(x_1, y_1)$, and, just as in the case of one autosomal locus, the effect $\boldsymbol{\alpha}(x_1, y_1)$ would be defined as

$$\boldsymbol{\alpha}(x_1, y_1) = \boldsymbol{\mu}(x_1, y_1) - \boldsymbol{\mu} - \boldsymbol{\alpha}(x_1) - \boldsymbol{\alpha}(y_1) \tag{5.25}$$

for all pairs of alleles $(x_1, y_1) \in \mathbb{A}_1 \times \mathbb{A}_1$. Observe that this effect is a measure of intralocus interactions of alleles. If the alleles at locus 1 acted in a purely additive manner, one would expect that $\boldsymbol{\alpha}(x_1, y_1)$ would be small, particularly if $x_1 = y_1$ were the same allele. But, if $x_1 \neq y_1$, then there may be intra-allelic interaction between the two alleles so that this effect may be greater than that for homozygous genotypes. If $A = \{x_1, x_2\}$, then the effect

$$\boldsymbol{\alpha}(x_1, x_2) = \boldsymbol{\mu}(x_1, x_2) - \boldsymbol{\mu} - \boldsymbol{\alpha}(x_1) - \boldsymbol{\alpha}(x_2) \tag{5.26}$$

is a measure of interloci or epistatic interactions for all pairs of alleles $(x_1, x_2) \in \mathbb{A}_1 \times \mathbb{A}_2$. It is of interest to note that among the four remaining effects for sets $A \in \mathfrak{T}_2$, three would be measures on interloci interactions and one would be an effect similar to that in (5.25) corresponding to positions $\{3, 4\}$.

For every set $A \in \mathfrak{T}_3$, there corresponds an effect $\boldsymbol{\alpha}(z(A))$ and, as can be seen from (5.19), there are four subsets in $\mathfrak{T}_3$. To illustrate the procedure used to define each of these effects, suppose that $A = \{1, 2, 3\}$. Then, $\boldsymbol{\mu}(x_1, y_1, x_2)$ is the conditional expectation that needs to be derived, using formula (5.24). Briefly, for each subset B of A, such that $A \neq B$, there will be an effect that is used in defining the effect $\boldsymbol{\alpha}(z(A))$. In particular,

$$\begin{aligned} \boldsymbol{\alpha}(x_1, y_1, x_2) &= \boldsymbol{\mu}(x_1, y_1, x_2) - \boldsymbol{\alpha}(x_1) - \boldsymbol{\alpha}(y_1) - \boldsymbol{\alpha}(x_2) \\ &\quad - \boldsymbol{\alpha}(x_1, y_1) - \boldsymbol{\alpha}(x_1, x_2) - \boldsymbol{\alpha}(y_1, x_2) \end{aligned} \tag{5.27}$$

for all triples $(x_1, y_1, x_2) \in \mathbb{A}_1 \times \mathbb{A}_1 \times \mathbb{A}_2$. The procedure illustrated in (5.27) may also be used to derive an expression for $\boldsymbol{\alpha}(z(A))$ for any set $A \in \mathfrak{T}_3$ such that $A \neq \{1, 2, 3\}$.

The last effect that needs to be defined is that for the class

$$\mathfrak{T}_4 = \{\mathfrak{G} = (1, 2, 3, 4)\}.$$

Let $\alpha(z)$, where $z = (x_1, y_1, x_2, y_2)$, denote this effect. Then, $\alpha(z)$ is determined by solving the equation

$$\boldsymbol{\mu}(z) = \boldsymbol{\mu} + \sum_{A \in \mathfrak{T}_1} \boldsymbol{\alpha}(z(A)) + \sum_{A \in \mathfrak{T}_2} \boldsymbol{\alpha}(z(A)) + \sum_{A \in \mathfrak{T}_3} \boldsymbol{\alpha}(z(A)) + \boldsymbol{\alpha}(z) \tag{5.28}$$

for $\alpha(z)$ for all genotypes $z \in \mathbb{G}$. The number of solutions to this equation depends on the number of alleles at each locus. For example, if there are two alleles at each locus, then there are 16 possible values of $\boldsymbol{\alpha}(z) = \boldsymbol{\alpha}((x_1, y_1, x_2, y_2))$. Moreover, each solution is a $k \times 1$ column vector, corresponding to the $k \geq 2$ traits under consideration. Let $\alpha_\nu(x_1, y_1, x_2, y_2)$ denote the effect for trait $\nu = 1, 2, \ldots, k$ in the $k \times 1$ vector $\boldsymbol{\alpha}(z)$. An investigator may be interested in estimating the component of variance

$$\text{var}[W_\nu] = \sum_{z \in \mathbb{G}} p(z) \alpha_\nu^2(z) \tag{5.29}$$

for each $\nu = 1, 2, \ldots, k$. But, because this is a summary statistic, it may mask the most interesting effects, i.e. those with the largest values, making up this variance component. Furthermore, because each effect in the sum (5.29) has been estimated, it would be possible to inspect the numerical values of each term in the sum of equation (5.29) for indications of usually large values that would be indicative of significant interactions among the alleles at the two loci under consideration. In the next section, a procedure for inspecting the numerical values of all the estimated effects will be suggested.

Before proceeding to the next section, it is interesting to note that Hemani *et al.* (2013) [10] have provided an evolutionary perspective on epistasis and the missing heritability and have also suggested that genome-wide association studies would be improved by searching directly for epistatic effects. It seems plausible, therefore, that the measures of epistatic effects defined in this section as well as those for multilocus effects that will be introduced in a subsequent section may provide not only a means of searching for epistatic effects in quantitative genetics but also for measuring these effects in genome-wide association studies.

6. Searching for Unusual Effects and Interactions Among the Alleles at Two Autosomal Loci

The strategy for searching for unusual effects and interactions at the two autosomal loci will be that of inspecting the squares of each effect and then picking out the largest of them as indicators of unusual effects of alleles or interactions among the alleles at the two loci. It will be assumed that $k \geq 2$ traits are under consideration, but to simplify the notation each set of effects will not be distinguished by the subscript ν, indicating the trait, but it will be tacitly assumed that any search procedure would be carried out for each trait. Let

$$\mathfrak{E}_1 = \left\{\alpha^2(z(A))|A \in \mathfrak{T}_1\right\} \tag{6.1}$$

denote the set of squared first order effects. For the case where there are two alleles at each locus, any position may be occupied by one of two alleles, and because there are four positions under consideration, there are eight effects in the set $\mathfrak{E}_1$. Consequently, in this case, an investigator could easily find the effect with the largest value by inspection, but in cases with a large number of effects it would be necessary to write a computer program or use existing software to find the largest of them.

The set of second order effects may be partitioned into two subsets; see (5.17) and (5.18). The set of squared effects for intra-allelic interactions is

$$\mathfrak{E}_{2IAI} = \left\{\alpha^2(z(A))|A \in \mathfrak{T}_{2IAI}\right\}. \tag{6.2}$$

In this case each of the two sets in $A \in \mathfrak{T}_{2IAI}$ contains two positions and each position may be occupied by two alleles. Hence, the number of effects in the set $\mathfrak{E}_{2IAI}$ is eight. The set of squared effects for second order epistatic interactions is

$$\mathfrak{E}_{2EPI} = \left\{\alpha^2(z(A))|A \in \mathfrak{T}_{2EPI}\right\}. \tag{6.3}$$

For this case, there are four sets of two positions in the set $\mathfrak{T}_{2EPI}$, and each position may be occupied by two positions. Consequently, the number of effects in the set $\mathfrak{E}_{2EPI}$ is 16.

The set of squared effects for third order epistatic interactions is

$$\mathfrak{E}_{3EPI} = \left\{\alpha^2(z(A))|A \in \mathfrak{T}_3\right\}. \tag{6.4}$$

Each of the four sets in $\mathfrak{T}_3$ consists of three positions and each position may be occupied with one of two alleles. Consequently, for each set of three positions there will correspond $2^3 = 8$ effects and, because there are four sets in $\mathfrak{T}_3$, there will be a total of 32 effects in the set $\mathfrak{E}_{3EPI}$. Rather than relying on a visual inspection of 32 effects, in this case, as well as in cases in which three or more autosomal loci are under consideration, an investigator may prefer to write a computer program or use existing software to find the largest squared effect. On the other hand, the investigator may not want to focus on the largest effect in each case just enumerated, but have a computer order the squared effects from the smallest to the largest so that, for example, attention could be focused on the three largest squared effects. For the case of two autosomal loci, the set of squared fourth order effects is

$$\mathfrak{E}_{4EPI} = \left\{\alpha^2(z(A))|A \in \mathfrak{T}_4 = \{1, 2, 3, 4\}\right\}. \tag{6.5}$$

As mentioned in Section 5, the number of squared effects in this set is 16 so that in this case an investigator may wish to select the largest one in a search for unusual epistatic interactions of fourth order.

If an investigator were interested in estimating components of the total genetic variance for any trait, it would be possible to do so for any of the sets of squared effects listed above. For example, suppose that attention was focused on the sum of the variance components corresponding to the squared effects in the set $\mathfrak{E}_{3\text{EPI}}$. For every $A \in \mathfrak{T}_3$, let $p(z(A))$ denote the corresponding marginal probability. Then, for the trait $\nu = 1, 2, \ldots, k$ under consideration, the component of the genetic variance corresponding to the set $\mathfrak{E}_{3\text{EPI}}$ is defined by

$$\text{var}_{\mathfrak{E}_{3\text{EPI}}}[W_\nu] = \sum_{A \in \mathfrak{T}_3} p(z(A))\alpha^2(z(A)), \tag{6.6}$$

and the variance components corresponding to the other sets listed above could be estimated similarly. On the other hand, one could proceed as in the one-locus case discussed in Section 5 and define a $15k \times 1$ column vector $\mathbf{\Phi}(x, y)$ as in (3.9) and a $15k \times 15k$ covariance matrix $\mathbf{\Psi}_G$ as in (3.11), but the formal details of this derivation will be left to the reader as an exercise. This covariance matrix may be particularly interesting when the sample or population is not in linkage equilibrium at the two loci under consideration.

At this point in the discussion, it is appropriate to mention the pioneering paper by Cockerham (1954) [7], who was among the first to consider variance components for epistatic interactions in quantitative genetics and introduced a distinctive nomenclature for such interactions. In his terminology but in a different notation, the genetic variance component

$$\text{var}_A[W_\nu] = \sum_{B \in \mathfrak{T}_1} p(z(B))\alpha^2(z(A)), \tag{6.7}$$

corresponding to the set $\mathfrak{T}_1$, would be called the additive component with subscript (A) for trait $\nu = 1, 2, \ldots, k$. The component of variance $\text{var}_D[W_\nu]$, corresponding to the set $\mathfrak{E}_{2\text{IAI}}$ would be called the dominance component with subscript (D), and the variance component $\text{var}_{AA}[W_\nu]$, corresponding to the set $\mathfrak{E}_{2\text{EPI}}$ would be called the additive by additive component with subscript (AA). Similarly, the variance component corresponding to the set $\mathfrak{E}_{3\text{EPI}}$ would be designated as the additive by dominant component, with subscript (AD). Finally, that corresponding to the set $\mathfrak{E}_{4\text{EPI}}$ would be labeled the dominant by dominant, with subscript (DD), component of the genetic variance for any trait $\nu = 1, 2, \ldots, k$. It should also be mentioned that equation (3.16), representing a partition of the genetic covariance matrix into component matrices for the case of one autosomal locus, could also generalized to the case of two autosomal loci using the set-based classification of effects outlined in this section, but this matrix partition will also be left to the reader as an exercise.

As will be discussed in the next section, when many loci are under consideration, the potential number of effects that may be considered can be very large. In such cases, to find unusually large effects and interactions even for the relatively simple case of only two alleles per locus may entail searches of hundreds of thousands or even millions of squared effects along with testing for their statistical significance. As has been widely recognized, when large numbers of statistical tests are considered, there is a risk of false discovery rates. It is beyond the scope of this paper to discuss the technicalities underlying false discovery rates, but it is suggested that the interested reader consult the papers Benjamini and Hochberg (1995) [2], Benjamini and Yekutieli (2001) [3] and Benjamini and Yekutieli (2005) [4].

7. An Overview of Cases for $l > 2$ Autosomal Loci

Let $l > 2$ denote the number of autosomal loci under consideration in some diploid species such as man. Then, the number of positions that may be occupied by alleles at each locus is $N = 2l$. Let $\mathfrak{S}$ be the set of N positions numbered $1, 2, \ldots, N$, let $\mathfrak{T}$ denote the class of all subsets of $\mathfrak{S}$ and let $\mathfrak{T}_\nu$ denote that class of subsets containing ν positions for $\nu = 0, 1, 2, \ldots, N$. Then, as is well known, the number of subsets in the class $\mathfrak{T}$ is 2^N and the number of subsets in the class $\mathfrak{T}_\nu$ is

$$\binom{N}{\nu} \tag{7.1}$$

for $\nu = 0, 1, 2, \ldots, N$. From the point of view of judging the feasibility of defining effects based on $\nu \geq 1$ positions, this number is essential, because it gives us the number of effects that may be defined. In what follows, it will also be helpful to recall the well-known equation from combinatorics,

$$\sum_{\nu=0}^{N} \binom{N}{\nu} = 2^N, \tag{7.2}$$

which is sometimes listed in textbooks on introductory probability theory. Because no effect is associated with the empty set φ, which is the only member of the class $\mathfrak{T}_0$ and $\binom{N}{0} = 1$, it follows that for the case of l autosomal loci the number of effects that may be defined is $2^N - 1$.

By way of an illustrative example, suppose that the number of autosomal loci under consideration is $l = 4$ so that $N = 8$. Thus, in this case it would be possible to define

$$2^8 - 1 = 255 \tag{7.3}$$

effects. It is also interesting to note that if there were two alleles per locus, then the number of possible genotypes would be $2^8 = 256$. If an investigator had access to a large sample of individuals whose genomes had been sequenced, it might be feasible to consider this many possible genotypes. But the study of 256 genotypes in small samples would be problematic, because even in a sample of size $n = 256$ all possible genotypes may be not be represented and in rare samples there may be only one observation for each genotype. If a sample were sufficiently large, an investigator might undertake a four-autosomal-locus study, but as a first approximation a decision may be made to define and estimate only first, second and third order effects rather than dealing with all the 255 possible effects.

It is clear that the number of sets in the class $\mathfrak{T}_1$ is eight so that eight first order effects could be defined and estimated, using the principles outlined in Section 5 for the case of two loci. The number of second order effects would be

$$\binom{8}{2} = 28 \tag{7.4}$$

and the number of third order effects would be

$$\binom{8}{3} = 56. \tag{7.5}$$

If an investigator were skilled in writing computer code, it would be possible to write programs to enumerate the sets in the classes $\mathfrak{T}_2$ and $\mathfrak{T}_3$ as well as to compute the effects associated with each class. Ideally, it would be close to optimal if a team of investigators were working on the project. At a minimum, a team should consist of at least two individuals: one individual should have expertise in genetics and managing databases consisting of individuals whose genomes have been sequenced, and a second individual should have expertise in writing software. It is interesting to note that with respect to the class $\mathfrak{T}_2$, there would be four sets of two positions such that effects that were measures of intra-allelic interactions could be estimated, but the remaining 24 sets would form a basis for defining and estimating effects that are measures of epistatic interactions among the alleles at four loci taken two at a time. All

56 effects corresponding to subsets in the class $\mathfrak{T}_3$ could be interpreted as measures of epistatic interactions among the alleles at the four loci under consideration. In order to include the phenomenon of pleiotropism in the formulation, it will be assumed, as in previous sections, that all effects are $k \times 1$ column vectors, where $k \geq 2$.

To provide a succinct overview of considering only first, second and third order effects in a variance component model, it will again be helpful to let z denote a genotype in the set $\mathbb{G}$ of all possible genotypes, and let $z(A)$ denote a set of alleles corresponding to the positions in any set $A \in \mathfrak{T}$. Next, suppose that all the effects in the classes

$$\{\boldsymbol{\alpha}(z(A)) | A \in \mathfrak{T}_\nu\} \tag{7.6}$$

have been defined and estimated for $\nu = 1, 2, 3$, using the principles outlined in Section 5. Then, the linear model expressing a genetic value $\boldsymbol{\mu}(z)$ as a function of effects would have the form

$$\boldsymbol{\mu}(z) = \boldsymbol{\mu} + \sum_{A \in \mathfrak{T}_1} \boldsymbol{\alpha}(z(A)) + \sum_{A \in \mathfrak{T}_2} \boldsymbol{\alpha}(z(A)) + \sum_{A \in \mathfrak{T}_3} \boldsymbol{\alpha}(z(A)) + \boldsymbol{\alpha}_R(z), \tag{7.7}$$

where the remainder effect is determined by solving equation (7.7) for $\boldsymbol{\alpha}_R(z)$ for every genotype $z \in \mathbb{G}$. For any trait ν in the vector-valued terms in this equation, an investigator could search for usually large squared effects following the search procedures suggested in Section 6 with the goal of finding epistatic interactions that would be interesting for interpreting the data. In this connection, any difference in these effects among the k traits under consideration would be attributable to pleiotropism.

To provide a measure of the adequacy of the approximation in (7.7) based on only first, second and third order effects would be of interest to estimate the variances for each trait in the vector of remainder effects $\boldsymbol{\alpha}_R(z)$. For example, let $\alpha_R^{(\nu)}(z)$ be the element in this vector for trait ν for $\nu = 1, 2, \ldots, k$ and let

$$\text{var}_R[W_\nu] = \sum_{z \in \mathbb{G}} p(z) \left(\alpha_R^{(\nu)}(z)\right)^2 \tag{7.8}$$

denote the estimated variance component corresponding to the remainder term in (7.7). Then, the ratio

$$\frac{\text{var}_R[W_\nu]}{\text{var}_G[W_\nu]}, \tag{7.9}$$

where $\text{var}_G[W_\nu] = \mathbf{cov}_{\nu\nu}[W_\nu]$ is element $\nu\nu$ on the principal diagonal of the genetic covariance matrix $\mathbf{cov}_G[W]$ in (2.23), is a measure of the contribution of the variance component in (7.8) to the total genetic variance for trait $\nu = 1, 2, \ldots, k$. In relative terms, small values of this fraction for each trait ν would be indicators of the goodness approximation in (7.7), using only first, second and third order effects.

At this point in the discussion of variance component models that are used in quantitative genetics, it is appropriate to mention that the number of genotypes under consideration may be significantly reduced if only three genotypes are identified at any locus for the case of two alleles per locus. To demonstrate this idea, it is helpful to resort to methods for representing genotypes used in classical Mendelian genetics. Suppose that at some locus there are two alleles B and b. Then, there are four genotypes BB, Bb, bB and aa. If, however, the heterozygotes Bb and bB are lumped into one class, then only three genotypes would be distinguishable at each locus. Thus, if this idea were used, the number of distinguishable genotypes with respect to four autosomal loci would be

$$3^4 = 81. \tag{7.10}$$

This number is significantly less than 256, which was the number of genotypes in which eight positions were used in defining genotypes. If this classification of genotypes were used, the procedures for estimating effects would need to be modified by taking into account the lumping of heterozygotes into one class.

During the past five to ten years, a rather large number of papers have been published by members of the genetic community in which genome-wide sweeps have been made of the human genome with goals of finding genomic regions that are implicated in such neurological conditions as Alzheimer's and Parkinson's disease. In an interesting paper, Raj *et al.* (2012) consider 11 regions (loci) in the human genome that have been implicated with Alzheimer's disease and present evidence that four of these loci are involved in a protein interaction network that has been maintained in the population by positive natural selection. It is also stated in this paper that 12 loci have been implicated in Parkinson's disease. It is of interest, therefore, to consider the extent to which the methodology presented in this paper may be applied to traits whose genetics are governed by 11 or 12 loci with two alleles per locus. If three genotypes were distinguished at each locus, then, for the case of 11 loci, the number of genotypes that could be distinguished for 11 loci would be

$$3^{11} = 177{,}147, \tag{7.11}$$

and for the case of 12 loci this number would be 531,441. When it is required that the genomes of all individuals in a sample have been sequenced, it is doubtful at the present time whether samples sizes of $n > 177{,}147$ or $n > 531{,}441$ would be available to an investigator or a team of investigators who are interested in applying the ideas presented in this paper.

Even though sample sizes of this magnitude may not be available to an investigator, their research, could proceed by using the available data to detect epistatic and pleiotropic interactions among the alleles represented in the sample. For example, for the case of 11 loci, it is suggested that the investigator perform a preliminary survey of the data to estimate the number $n(x, y)$ of individuals of genotype (x, y) are available, where the alleles in x and y have been ascertained with respect to 11 or fewer loci. If some number or numbers $n(x, y)$ are too small to yield reliable statistical information, then the investigator may make a decision to restrict attention to a subset of the 11 loci such that the sample sizes for each genotype are judged sufficiently large for one to draw statistically reliable inferences from the data as to the presence of epistatic and pleiotropic interactions.

ACKNOWLEDGMENTS

A word of thanks is due to Dr. Towfique Raj, Division of Genetics, Brigham and Women's Hospital, Harvard Medical School, Boston, MA 02115, USA, who called the author's attention to recent papers devoted to the estimation of heritability of various quantitative traits in humans, which have been cited in this paper. A cooperative research effort involving the author and Dr. Raj's group is also in progress, with a goal of writing software to implement some of the ideas set forth in this paper, along with results not included in this paper, and applying them in a quantitative genetic analysis of data from samples of patients whose genomes have been sequenced.

REFERENCES

1. Anderson, T. W. (1984) *An Introduction to Multivariate Statistical Analysis*, 2nd ed. John Wiley and Sons, New York, Chichester, Brisbane, Singapore.
2. Benjamini, Y. and Hochberg, Y. (1995) Controlling false discovery rate: a practical and powerful approach to multiple testing. *J. R. Soc. Ser. B* **57**: 289–300.
3. Benjamini, Y. and Yekutieli, D. (2001) The control of false discovery rate under dependency. *Ann. Stat.* **29**: 1165–1188.
4. Benjamini, Y. and Yekutieli, D. (2005) False discovery rate adjusted multiple confidence intervals for selected parameters. *J. Am. Stat. Assoc.* **100**: 71–93.
5. Bulmer, M. G. (1980) *The Mathematical Theory of Quantitative Genetics*. Clarendon, Oxford.
6. Church, G. M. (2006) Genomes for all. *Sci. Am.* **294**: 46–54.
7. Cockerham, C. C. (1954) An extension of the concept of partitioning the hereditary variance for analysis of covariance among relatives when epistasis is present. *Genetics* **39**: 859–882.
8. Falconer, D. and MacKay, T. F. C. (1996) *Introduction to Quantitative Genetics*. Longman, New York.
9. Fisher, R. A. (1918) The correlation among relatives on the assumption of Mendelian inheritance. *Trans. R. Soc., Edinb.* **52**: 399–433.
10. Hemani, G., Knott, S. and Haley, C. (2013) An evolutionary perspective on epistasis and the missing heritability. *PLoS Genet.* **9(2)**: e1003295. DOI: 10.1371/journal.pgen.1003295.
11. Kao, C.-H. and Zeng, Z.-B. (2002) Modeling epistasis of quantitative trait loci using Cockerham model. *Genetics* **160**: 1243–1261.
12. Kempthorne, O. (1954) The correlations between relatives in a random mating population. *Proc. R. Soc. London, B* **143**: 103–113.
13. Kempthorne, O. (1957) *An Introduction to Genetic Statistics*. John Wiley and Sons, New York.
14. Laird, N. M. and Lange, C. (2011) *The Fundamentals of Modern Statistical Genetics*. DOI: 10.1007/978-1-4419-7338-2. Springer, New York, Dordrecht, Heidelberg, London.
15. Liu, B. H. (1998) *Statistical Genomics Linkage, Mapping and QTL Analysis*. CRC, Boca Raton, London, New York Washington, D.C.
16. Lynch, M. and Walsh, B. (1998) *Genetics and the Analysis of Quantitative Traits*. Sinauer Associates, Sunderland, MA, USA.
17. Mao, Y., Nicole, R., Ma, L., Dvorkin, D. and Da, Y. (2006) Detection of SNP epistasis effects of quantitative traits using extended Kempthorne model. *Physiol. Genomics* **28**: 46–52.

18. Mode, C. J. and Robinson, H. F. (1959) Pleiotropism and the genetic variance and covariance. *Biometrics* **15**: 518–537.
19. Muirhed, R. J. (1982) *Aspects of Multivariate Statistical Theory.* John Wiley and Sons, New York, Chichester, Brisbane, Singapore.
20. Price, A. L., Helgason, A., *et al.* (2011) Single tissue and cross-tissue heritability of gene expression via identity-by-descent in related and unrelated individuals. *PloS Genet.* 7(2) e1001317. DOI: 10.1371/journal.pgen.1001317.
21. Raj, T., Shulman, J. M., Keenan, B. T. Lori, B. Chibnik, L. B., Evans, D. A., Bennett, D. A., Stranger, B. E. and De Jager, P. L. (2012) Alzheimer disease susceptibility loci: evidence for a protein network under natural selection. DOI: 10.1016/j.ajhg.2012.02.022. _2012 The American Society of Human Genetics.
22. Rossin, E. J., Lage, K., Raychaudhuri, S., Xavier, R. J., Tatar, D., *et al.* (2011) Proteins encoded in genomic regions associated with immune-mediated disease physically interact and suggest underlying biology. *PLoS Genet.* **7**(**1**): e1001273. DOI: 10.1371/journal.pgen.1001273.
23. Stranger, B. E., Eli, A., Stahl, E. A. and Raj, T. (2010) Progress and promise of genome-wide association studies for human complex trait genetics. DOI: 10.1534/genetics.110.120907.
24. Wu, R.-L., Ma, C.-X., and Casella, G. (2010) *Statistical Genetics of Quantitative Traits, Linkage, Maps and QTL.* Springer Science, Berlin, New York.
25. Yang, J., Benyamin, B., *et al.* (2010) Common SNPs explain a large proportion of heritability for human height. *Nat. Genet.* ; **42**(7): 565–569. DOI: 10.1038/ng.608.
26. Zaitlen, N., Kraft, P., *et al.* (2013) Using extended genealogy to estimate components of heritability for 23 quantitative and dichotomous traits. *PLoS Genet.* **9**(**5**): e1003520. DOI: 10.1371/journal.pgen.1003520.

CHAPTER 5

Direct Estimation of Effects and Tests of Their Statistical Significance for the Case of One Autosomal Locus with Two Alleles

ABSTRACT

In previous papers (see the references), the author introduced methods for estimating effects directly in samples of individuals whose genomes had been sequenced for the cases of one and two or more quantitative traits. In those papers, no attention was given to developing procedures for testing the statistical significance of the estimated effects. This paper is devoted to the development of statistical tests of significance of estimated effects for the simple case of one autosomal locus with two alleles, using Monte Carlo simulation methods. Because no real data was available to the author, artificial data for the three genotypes was simulated by using Monte Carlo simulation procedures with fixed sample size for each genotype as well as expectations and variances. In all cases considered, the null hypothesis was described in detail so as to inform the reader about the basic concepts underlying the proposed tests of statistical significance. For the class of statistical tests described in this paper, two types of p values may be distinguished. One type of p values consists of those for each of the estimated effects. A second type of p values concerns the joint statistical significance of two or more estimated effects. The consideration of the simple case of two autosomal loci is useful, because it provides insights into how the Monte Carlo simulation procedures used in this paper may be extended to cases of two or more autosomal loci with two or more alleles at each locus.

*This chapter was published as a paper in an online journal: Mode, C. J. (2015) Direct estimation of effects and tests of their statistical significance for the case of one autosomal locus with two alleles. Version 1.0. *Global Journal of Science Frontier Research — Mathematics and Decision Sciences* (2015). Online ISSN: 2249–4626. Print ISSN: 0975-5896. The content of this chapter is a PDF image of the LaTeX document of the paper that was accepted for publication.

1. Introduction

As was suggested in previous papers, when an investigator is dealing with a sample of individuals whose genomes have been sequenced and a set of regions of the genome have been identified that affect the expression of a quantitative trait or traits, then it becomes possible to provide a working definition of a set of loci in each individual at the genomic level; see Chapters 3 and 4. Moreover, if it is also possible to use markers in the DNA of each individual to provide working definitions of at least two alleles at each locus, then the investigator can develop a concrete working definition of the set of loci with two alleles at each that have been shown to have an effect on the expression of a quantitative trait or traits as expressed in a numerical measurement or measurements on each individual in the sample.

In principle, a quantitative trait or traits may be analyzed statistically for any combination of the set of loci under consideration, but because the number of genotypes that can be identified increases at a fast rate as the number of loci under consideration increases, the sample of individuals may not be sufficiently large to ensure that the number of individuals of each genotype is large enough for one to obtain statistically significant results, as discussed in Chapters 3 and 4. Such a situation will usually arise whenever the number of loci under consideration is greater than five or six. Therefore, if an investigator has six or more loci under consideration in a sample of data, it would be prudent to perform a preliminary analysis of the data for each locus under consideration in order to develop an understanding as to which combination of loci would be most informative and fruitful to explore.

To execute such an experiment, an investigator would need software to estimate each effect whose square is a component of the genetic variance for each locus under consideration. In the papers presented in Chapters 3 and 4, it was assumed that any allele in a genotype could be identified as to whether it was contributed by the father or mother of any individual in the sample. In many data sets, however, this assumption is not valid so that the investigator could identify only three genotypes per locus for the case of two alleles per locus. These three genotypes consist of two homozygotes and a heterozygote for which it was not possible to identify whether each allele was maternal or paternal in origin. The purpose of this chapter is to provide an overview of the software necessary to carry out a preliminary exploration of a data set, such that the genome of each individual in the set has been sequenced, for each locus under consideration. The main focus of the chapter is to implement software to do the necessary computations for the simplest case of one autosomal locus with two alleles with a view to extending the software to cases of two or more autosomal loci. A mathematical description of this software is provided for those investigators who write code using a programing language based on the manipulation of arrays such as APL or MATLAB.

A Monte Carlo simulation procedure was used to provide data for illustrating how the software may be used to analyze real data when it is available. Contained in this software are procedures for estimating the squares of effects and a description of a procedure for formulating null hypotheses to test the statistical significance of the estimated squares of effects. After a null hypothesis is defined, a Monte Carlo simulation procedure was used to estimate p values to judge whether each estimated square of an effect was statistically significant, given some null hypothesis. Detailed technical descriptions of null hypotheses that were used in tests of statistical significance are also provided for each reported experiment.

2. Nonidentifiability of Heterozygotes in the Case of One Autosomal Locus

In previous work on defining effects in connection with components of variance models in quantitative genetics, it was assumed that two kinds of heterozygotes could be identified when one was working with a sample of individuals whose genomes have been sequenced. For example, let the symbol 1 denote the presence of a marker on a haplotype with respect to some locus, and let the symbol 0 denote the absence of this marker. Then, if it is assumed that it is possible to detect in each individual whether each of the two alleles in a genotype was contributed by the maternal or paternal parent, a genotype could be represented by the symbol (x, y), where x and y denote, respectively, the maternal and the paternal allele. Thus, if both x and y are assigned the symbol 1 or 0, then four genotypes in the set

$$\mathbb{G}_1 = \{(1,1), (1,0), (0,1), (0,0)\} \tag{2.1}$$

can be identified in a sample of individuals. But if it is not possible to identify the maternal and paternal alleles in an individual, then the two possible heterozygotes, $(1, 0)$ and $(0, 1)$, would be lumped into a single category called heterozygotes.

However, in such samples, the two homozygotes, $(1, 1)$ and $(0, 0)$, could be identified unambiguously. In what follows, the genotypes in the three categories will be denoted by the symbols $(1, 1)$, $(x \neq y)$ and $(0, 0)$, where $(x \neq y)$ stands for heterozygotes in the nonidentifiable case. In this case,

$$\mathbb{G}_2 = \{(1, 1), (x \neq y), (0, 0)\} = \{g_1, g_2, g_3\} \tag{2.2}$$

is the set of genotypes of recognizable genotypes. To simplify the notation in this case, the symbols g_1, g_2, g_3 will stand for three genotypes under consideration, as indicated in (2.2).

For the case of nonidentifible heterozygotes, let $n > 1$ denote the number of individuals in a sample, and let $n(1, 1)$, $n(x \neq y)$ and $n(0, 0)$ denote, respectively, the number of individuals of each of the three genotypes. To simplify the notation, the number of each of these three genotypes indicated in the set

$$\{n(g_1), n(g_2), n(g_3)\} \tag{2.3}$$

denotes the set of genotypes, and let g, with or without subscripts, denote an element in the set $\mathbb{G}_2$ as indicated on the right in (2.2). Then,

$$n = \sum_{g \in \mathbb{G}_2} n(g) \tag{2.4}$$

is the number of individuals in the sample, and

$$p_\nu = \frac{n(g_\nu)}{n} \tag{2.5}$$

is the estimated frequency of genotype g_ν in the population for $\nu = 1, 2, 3$. This collection of estimates will be referred to as the estimated genotypic distribution.

To take into account the problem of setting up a structure that incorporates phenotypic variation among individuals of the same genotype in a sample with respect to some quantitative trait, let W denote a random variable taking values in the set $\mathbb{R}_W$ of possible values of the phenotype. Usually, $\mathbb{R}_W$ is a set of rational real numbers. Given genotype $g \in \mathbb{G}_2$, let $f(w|g)$ denote the conditional probability density function of the random phenotypic variable W, and suppose that there are $n(g)$ observed realizations of the random variable W denoted by the symbols W_i, for $i = 1, 2, \ldots, n(g)$. The conditional expectation of the random variable W, given the genotype $g \in \mathbb{G}$, is

$$E[W|g] = \int_{\mathbb{R}_W} f(w|g)dw \tag{2.6}$$

for every genotype $g \in \mathbb{G}_2$.

Therefore, an estimator of the conditional expectation $E[W|g]$ is

$$\mu(g) = \frac{1}{n(g)} \sum_{i=1}^{n(g)} W_i \tag{2.7}$$

for all $g \in \mathbb{G}$. In what follows, these estimates will be referred to as the genetic values. Let μ denote the unconditional expectation of the genetic values. Then,

$$\mu = \sum_{g \in \mathbb{G}_2} p(g)\mu(g). \tag{2.8}$$

Observe that a formula of this type would also be valid if the set $\mathbb{G}_1$ of genotypes were under consideration; in this case there would be four genotypes to take into account.

One objective of this chapter is to suggest ways of writing software to implement the formulas above, using an array-manipulating programming language. Included in the class of array-manipulating programming languages are APL and MATLAB. For such languages, a good starting point is to represent the genotypic distribution in (2.5) and the set of estimated expectations in (2.7) in matrix form. To cast the genotypic distribution in matrix form, suppose that the set of genotypes under consideration is $\mathbb{G}_2$. Then, the matrix of estimated genetic values will have the form

$$\boldsymbol{\mu}_{\mathbb{G}_2} = \begin{bmatrix} \mu(g_1) & \mu(g_2) \\ 0 & \mu(g_3) \end{bmatrix}. \tag{2.9}$$

If both of the genotypes $(1,0)$ and $(0,1)$, i.e. both the maternal and paternal alleles in a sample of individuals, can be identified, then the set of genotypes under consideration would be the set $\mathbb{G}_1$ as defined above. In this case, the matrix of genetic values has the form

$$\boldsymbol{\mu}_{\mathbb{G}_1} = \begin{bmatrix} \mu(1,1) & \mu(1,0) \\ \mu(0,1) & \mu(0,0) \end{bmatrix} = [\mu(i,j)], \tag{2.10}$$

where in this case $\mu(0,1)$ may be positive.

Let the array

$$\boldsymbol{p}_{\mathbb{G}_1} = \begin{bmatrix} p(1,1) & p(1,0) \\ p(0,1) & p(0,0) \end{bmatrix} = [p(i,j)] \tag{2.11}$$

represent the matrix form of the genotypic distribution for case 1. For the case where the set of genotypes $\mathbb{G}_1$ is under consideration, let a matrix operation of element-by-element multiplication be denoted by

$$\boldsymbol{p} \times \boldsymbol{\mu} = [p(i,j)\mu(i,j)]. \tag{2.12}$$

The symbol on the left in (2.12) would be the format for doing the matrix operation of element-by-element multiplication in APL. Then, for case 1, the unconditional expectation of the matrix of genetic values may be written in the form

$$\mu_{\mathbb{G}_1} = \sum_{(i,j)\in\mathbb{G}_1} [p(i,j)\mu(i,j)]. \tag{2.13}$$

It is important to observe that if the set $\mathbb{G}_2$ of genotypes were under consideration, then a formula of the type (2.13) could also be used to compute unconditional expectation $\mu_{\mathbb{G}_2}$. From the point of view of writing software with an array-processing programming language, it is important to observe that the same program could be used for these cases except case 2, where the matrix of expected genetic values would have the form in (2.9) and the matrix form of the genotypic distribution would be represented in the form

$$p_{\mathbb{G}_2} = \begin{bmatrix} p(g_1) & p(g_2) \\ 0 & p(g_3) \end{bmatrix}. \tag{2.14}$$

In what follows in this section, a general notation will be used to partition the phenotypic variance of a trait into the genetic and environmental variances, using a general notation that includes cases 1 and 2 described above. Let $\mathbb{A}$ denote the set of alleles at some autosomal locus, let (x, y) be any genotype, where $x \in \mathbb{A}$ and $y \in \mathbb{A}$, and let

$$\mathbb{G} = \{(x, y) | x \in \mathbb{A} \text{ and } y \in \mathbb{A}\} \tag{2.15}$$

be the set of genotypes under consideration. Given genotype $(x, y) \in \mathbb{G}$, a set of genotypes under consideration, let $W(x, y)$ denote a realization of the phenotypic random variable W, given (x, y). Then observe that the equation

$$W(x, y) - \mu = (\mu(x, y) - \mu) + (W(x, y) - \mu(x, y)) \tag{2.16}$$

is valid for all genotypes $(x, y) \in \mathbb{G}$.

The phenotypic variance, genetic and environmental variances are defined as

$$\mathrm{var}_P[W] = \sum_{(x,y)} p(x, y)(W(x, y) - \mu)^2, \tag{2.17}$$

$$\mathrm{var}_G[W] = \sum_{(x,y)} p(x, y)(\mu(x, y) - \mu)^2, \tag{2.18}$$

$$\mathrm{var}_E[W] = \sum_{(x,y)} p(x, y) \left(W(x, y) - \mu(x, y)\right)^2, \tag{2.19}$$

respectively.

By using a conditioning argument given (x, y) and (2.16), it can be shown that the equation

$$\mathrm{var}_P[W] = \mathrm{var}_G[W] + \mathrm{var}_E[W] \tag{2.20}$$

is valid. If the reader is interested in a detailed derivation of the formula in (2.20), consult Chapter 3, which contains a detailed account for the case of one quantitative trait. Observe that if the variances on the right of equation

(2.20) are estimates based on data, then the sum on the right of (2.20) is an estimator of $\text{var}_P[W]$, the phenotypic variance. By definition, H, the heritability of a trait, is

$$H = \frac{\text{var}_G[W]}{\text{var}_G[W] + \text{var}_E[W]} = \frac{\text{var}_G[W]}{\text{var}_P[W]}. \tag{2.21}$$

From this formula, it can be seen that to estimate H, it suffices to compute $\text{var}_G[W]$ and $\text{var}_E[W]$, and then use equation (2.20) to check the validity of formula (2.20) numerically. Note that any estimate of H will satisfy the condition $0 \leq H \leq 1$. It should also be observed that from formula (2.17) the phenotypic variance $\text{var}_P[W]$ may also be computed directly.

The formulas just derived are theoretical and when a sample of data is available to an investigator, the genotypic distribution may be estimated as well as means and variances. For example, for each genotype $(i,j) \in \mathbb{G}_2$, let $W_\nu(i,j)$, for $\nu = 1, 2, \ldots, n(i,j)$, be a set of quantitative observations on some trait. Then, for every genotype $(i,j) \in \mathbb{G}_2$,

$$\widehat{\mu}(i,j) = \frac{1}{n(i,j)} \sum_{\nu=1}^{n(i,i)} W(i,j) \tag{2.22}$$

is an estimate of the expectation $\mu(i,j)$, and an estimate of $\sigma^2(i,j)$ is

$$\widehat{\sigma^2}(i,j) = \frac{1}{n(i,j) - 1} \sum_{\nu=1}^{n(i,j)} \left(W(i,j) - \widehat{\mu}(i,j)\right)^2 \tag{2.23}$$

is an estimate of $\sigma^2(i,j)$ for $n(i,j) > 1$, and

$$\widehat{p}(i,j) = \frac{n(i,j)}{n} \tag{2.24}$$

is an estimate of the genotypic distribution. In the sections that follow, the symbol $\widehat{\circ}$ will be dropped to lighten the notation, but it will be tacitly understood that when a set of data is under consideration, the symbols $\mu(i,j), \sigma^2(i,j)$ and $p(i,j)$ are estimates of parameters.

3. Estimating Effects Directly for the Case of One Autosomal Locus

In classical quantitative genetics, the variances in variance component models were usually estimated indirectly by using analysis of variance–covariance procedures, but within this framework it was not possible to estimate the effects, because they were squared terms in the weighted sums that were, by definition, variance components. See Chapter 1 for a specific example of an analysis of variance–covarince procedure applied to data. But, as will be shown in this section, when the genotype of each individual in a sample is known at the DNA level, it is possible to estimate effects directly from the data for cases of at least two alleles at the locus, as was suggested in Chapters 3 and 4.

Suppose that the maternal and paternal alleles are not identified in a sample, and let $n(1,1)$, $n(0,0)$ and $n(\text{het})$ denote, respectively, the total number of individuals that were homozygous for the allele 1 or 0 and heterozygous for these alleles. Then, the total number of individuals in the sample is

$$n = n(1,1) + n(0,0) + n(\text{het}). \tag{3.1}$$

Let $p(1,1), p(0,0)$ and $p(\text{het})$ denote, respectively, the frequencies of the three genotypes under consideration. Then, $p(1,1) = n(1,1)/n$, $p(0,0) = n(0,0)/n$ and $p(\text{het}) = n(\text{het})/n$ are estimators of these frequencies. It is also essential to have estimates of the frequencies of alleles in the sample. Let $p(1)$ and $p(0)$ denote the estimated frequencies of alleles 1 and 0 in the sample. The number of copies of allele 1 in individuals of genotype $(1,1)$ is two, and the number of copies of this allele in each heterozygote is one. Therefore, the estimated frequency of allele 1 in the sample is

$$p(1) = \frac{2n(1,1) + n(\text{het})}{2n} = p(1,1) + \frac{1}{2}p(\text{het}). \tag{3.2}$$

Similarly, the estimate of allele 0 in the sample is

$$p(0) = p(0,0) + \frac{1}{2}p(\text{het}). \tag{3.3}$$

Observe that

$$p(1) + p(0) = 1, \tag{3.4}$$

as they should.

In order to define all effects for the case of one autosomal locus, it will be necessary to define conditional expectations of the estimated means defined in equation (2.7) in Section 2 with respect to the genotypic distribution defined below. When one is programming in an array-processing language, it is convenient to represent the data and estimates in matrix form. Let $p(1,0) = p(0,1) = 0.5p(\text{het})$. Then, the matrix form of the genotypic distribution is

$$\mathbf{p}_{\mathbb{G}_2} = \begin{bmatrix} p(1,1) & p(1,0) \\ p(0,1) & p(0,0) \end{bmatrix}. \tag{3.5}$$

The subscript $\mathbb{G}_2$ denotes the set of genotypes defined in (2.2) in Section 2. Given this matrix, the frequency of allele 1 has the form

$$p(1) = p(1,1) + p(1,0). \tag{3.6}$$

Similarly, the frequency of allele 0 has the form

$$p(0) = p(0,1) + p(0,0). \tag{3.7}$$

From the point of view of writing computer code in an array-processing language, one could use a single command to compute the sum of the rows of the matrix $\mathbf{p}_{\mathbb{G}_2}$ that would result in an array with the elements $p(1)$ and $p(0)$.

To expedite the writing of code in an array-processing programming language, let $\mu(1,1)$, $\mu(0,0)$ and $\mu(\text{het})$ denote the genetic means for three genotypes under consideration, and, to cast these means in matrix form as in (3.5), let $\mu(0,1) = \mu(1,0) = \mu(\text{het})$. Then, the matrix of these means may be represented in the form

$$\boldsymbol{\mu}_{\mathbb{G}_2} = \begin{bmatrix} \mu(1,1) & \mu(1,0) \\ \mu(0,1) & \mu(0,0) \end{bmatrix}. \tag{3.8}$$

An essential step in defining the effects in what follows is to estimate the mean μ defined in (2.8) of Section 2. To this end consider the matrix product

$$\mathbf{p}_{\mathbb{G}_2} \times \boldsymbol{\mu}_{\mathbb{G}_2}, \tag{3.9}$$

where the symbol $\times$ stands for element-by-element multiplication. Then, in this notation, the mean μ has the form

$$\mu = \sum \mathbf{p}_{\mathbb{G}_2} \times \boldsymbol{\mu}_{\mathbb{G}_2} = \sum_{(i,j)\in\mathbb{G}_2} p(i,j)\mu(i,j); \tag{3.10}$$

see (2.8) of Section 2. When one is writing code in an array-processing programing language, only a few symbols would be required to write the code to do the operations defined in (3.10).

Another essential step in defining effects is to define the conditional distributions based in terms of the elements of the matrix $\mathbf{p}_{G_2}$ in (3.5). By definition, the conditional distribution of the genotypic distribution, given allele 1, is

$$\frac{1}{p(1)}(p(1,1), p(1,0)). \tag{3.11}$$

Let $\mu(1)$ denote the conditional expectation of the means $(\mu(1,1)), \mu(1,0)$, given allele 1. Then, by definition,

$$\mu(1) = \frac{1}{p(1)}(p(1,1)\mu(1,1)) + p(1,0)\mu(1,0). \tag{3.12}$$

The conditional expectation $\mu(0)$ is defined similarly. Observe that if the sample is in a Hardy–Weinberg equilibrium so that $p(i,j) = p(i)p(j)$ for all genotypes $(i,j) \in \mathbb{G}_2$, then $\mu(1)$ has the form

$$\mu(1) = \frac{1}{p(1)}p^2(1)\mu(1,1) + p(1)p(0)\mu(1,0) = p(1)\mu(1,1) + p(0)\mu(1,0). \tag{3.13}$$

The conditional expectation $\mu(0)$ has a similar form when the sample is in a Hardy–Weinberg equilibrium.

Given the above definitions, the effect of allele 1 is defined by

$$\alpha(1) = \mu(1) - \mu. \tag{3.14}$$

Similarly, the effect of allele 0 is defined by

$$\alpha(0) = \mu(0) - \mu. \tag{3.15}$$

In what follows, the effects defined in (3.14) and (3.15) will be referred to as first order effects. Observe that the expectation of these effects with respect to the genotypic distribution is

$$E_{\mathbb{G}_2}[\alpha] = p(1)\alpha(1) + p(0)\alpha(0) = 0. \tag{3.16}$$

It is clear that the effects just defined can be estimated directly from the data, given that the genotype of every individual in the sample can be identified.

For reasons that will be made clear subsequently, the additive variance is defined by

$$\text{var}_A[W] = p(1)\alpha^2(1) + p(0)\alpha^2(0). \tag{3.17}$$

The symbol W has been included in the left side of this equation as a reminder that effects on the right side of the equation have been estimated from data. From a perspective of using an analysis of variance procedure to estimate the additive variance in (3.17), it would be impossible to estimate the effects defined in (3.14) and (3.15) from an estimate of $\text{var}_A[W]$. This observation clearly differentiates classical methods of estimating variance components, based on some analysis of variance procedure, from the direct method of estimating effects from the data as outlined above.

Additional effects can also be defined as measures of the presence of interactions among the two alleles in the genotype of any individual in the sample. For any genotype (i, j), a second order effect $\alpha(i, j)$ is defined as

$$\alpha(i,j) = \mu(i,j) - \mu - \alpha(i) - \alpha(j) \tag{3.18}$$

for all genotypes $(i, j) \in \mathbb{G}_2$. From this equation, it follows that

$$\mu(i,j) = \mu + \alpha(i) + \alpha(j) + \alpha(i,j) \tag{3.19}$$

for all genotypes $(i, j) \in \mathbb{G}_2$. If there are no interactions among alleles i and j, then $\alpha(i, j) = 0$ and equation (3.19) reduces to

$$\mu(i,j) = \mu + \alpha(i) + \alpha(j). \tag{3.20}$$

If this equation holds, then it is said that alleles i and j act additively, but if $\alpha(i, j) \neq 0$, then there is some interaction among alleles i and j.

If a sample of individuals is in a Hardy–Weinberg equilibrium (see definition below), then in the additive case the genetic variance has the form

$$\begin{aligned} \text{var}_G[W] &= \sum_{(i,j)\in\mathbb{G}_2} p(i,j)(\mu(i,j) - \mu)^2 \\ &= p(1)\alpha^2(1) + p(0)\alpha^2(0) = \text{var}_A[W], \end{aligned} \tag{3.21}$$

which justifies equation (3.17). By definition

$$\text{var}_{\text{IAI}}[W] = \sum_{(i,j)\in G_2} p(i,j)\alpha^2(i,j) \tag{3.22}$$

is the intra-allelic interaction variance. If the sample is in a Hardy–Weinberg equilibrium, then, by definition, $p(i, j) = p(i)p(j)$ for all alleles i and j, and it can be shown that

$$\text{var}_G[W] = \text{var}_A[W] + \text{var}_{\text{IAI}}[W]. \tag{3.23}$$

A proof of these results may be found in Chapter 3.

If the population is not in a Hardy–Weinberg equilibrium, then from equation, it can be shown that $\text{var}_G[W]$ would also contain covariance terms, but the details will be omitted; if the reader is interested in more details, Chapter 3 may again be consulted. In classical quantitative genetics, the objective of an experiment would be the estimation of the variance components on the left side of equation (3.23), using some analysis of variance procedure. But, as was shown above, when the genotype of every individual in the sample is known, all second order effects defined above can be estimated directly from the data.

For example, from equation (3.19) it follows that the formula for estimating the second order effect for genotype $(1, 1)$ is

$$\alpha(1,1) = \mu(1,1) - \mu - 2\alpha(1). \tag{3.24}$$

A similar formula for an estimator of the effect $\alpha(0,0)$ would also have the form of equation (3.24). The formula for estimating the second order effect for genotype (i, j) such that $(i \neq j)$ directly has the form

$$\alpha(1,2) = \mu(1,2) - \mu - \alpha(1) - \alpha(2). \tag{3.25}$$

As can be seen from the symmetric matrices in (3.5) and (3.8), it follows that $\alpha(2,1) = \alpha(1,2)$.

For the case in which maternal and paternal alleles can be identified in any genotype, the set of genotypes $\mathbb{G}_1 = \{(1,1),(1,0),(0,1),(0,0)\}$ (see Section 2 for more detailed comments), would be under consideration. In this case the 2×2 matrix $\boldsymbol{P}_{\mathbb{G}_1}$ would not be symmetric, because the relation $p(1,2) \neq p(2,1)$ would hold except for rare coincidences. Similarly, the 2×2 matrix $\boldsymbol{\mu}_{G_1}$ genetic values, conditional means, would be nonsymmetric. In this case it would also be necessary to distinguish the frequencies of maternal and paternal alleles. For example, if the rows of the matrix $\boldsymbol{P}_{\mathbb{G}_1}$ were summed, the result would be the distribution of maternal alleles $(p_{\text{mat}}(1), p_{\text{mat}}(0))$. Similarly, if the columns of this matrix were summed, the result would be the distribution of $(p_{\text{pat}}(1), p_{\text{pat}}(0))$ of paternal alleles. Given these allelic distributions, to write the computer code to compute the effects in this case would require only a few changes in the code for the case in which the maternal and the paternal cannot be distinguished as outlined above.

A question that naturally arises at this point in formulating the model under consideration is: How can we construct procedures to test the statistical significance of the estimated effects? Because the squares of the effects are summed when one is defining variance components, it seems fitting to use squared effects in designing tests of statistical significance. Another advantage of using squared effects is that their signs are always nonnegative. For the case in which the maternal and paternal alleles cannot be distinguished, let the 5×1 column vector

$$E = \begin{bmatrix} \alpha^2(1) \\ \alpha^2(0) \\ \alpha^2(1,1) \\ \alpha^2(0,0) \\ \alpha^2(1,0) \end{bmatrix} \tag{3.26}$$

denote the squared effects. For the case where the maternal and paternal alleles can be distinguished, this vector would contain eight elements. In the next section, procedures for testing the statistical significance of the elements in the vector in (3.26) will be presented.

4. Permutation Tests for Statistical Significance of Estimated Effects

In this section, a class of tests of statistical significance based on computer-intensive methods will be discussed. In the statistical literature, the class of tests described in this section are often referred to as permutation tests. For example, consider the case in which the maternal and paternal alleles cannot be distinguished. Then, the set of genotypes in the model would be $\mathbb{G}_2 = \{(1,1),(0,0),(1,0)\}$, where the symbol $(1,0)$ represents heterozygotes. Let $n(1,1)$, $n(0,0)$ and $n(1,0)$ denote the number of individuals of the three genotypes in a sample of

$$n = n(1,1) + n(0,0) + n(1,0) \tag{4.1}$$

individuals whose genomes have been sequenced. For each genotype $(i,j) \in \mathbb{G}_2$, let $W(i,j) = \{W_\nu(i,j) | \nu = 1,2,\ldots,n(i,j)\}$ denote the sample of $n(i,j)$ realizations of the phenotypic random variable W describing the variability in the expression of some quantitative trait for individuals of genotype (i,j). The combined data set under consideration may be represented as

$$\text{DATA} = W(1,1), W(0,0), W(1,0) \tag{4.2}$$

and consists of n observations.

The first step in setting up a permutation test of the data is to compute a random permutation, denoted by PERM, of the data set in (4.2). Given this random permutation of the data, the next step consists in choosing the first $n(1,1)$ elements of PERM as a sample of the quasi-observations of the $n(1,1)$ individuals of genotype $(1,1)$. Similarly, the next $n(0,0)$ elements of PERM would represent quasi-observations on the $n(0,0)$ of genotype $(0,0)$. Finally, the last $n(1,0)$ elements of PERM would represent the $n(1,0)$ quasi-observations on individuals of genotype $n(1,0)$. Then, suppose that these operations are repeated N times to generate a set of random mutations of the data in (4.2).

As is well-known, the set $\mathbb{S}$ of all permutations of the data chosen in this manner contains

$$M = \frac{n!}{n(1,1)!n(0,0)!n(1,0)!} \tag{4.3}$$

elements. For example, for the case $n(1,1) = 33$, $n(0,0) = 33$ and $n(1,0) = 34$, this number is

$$M = \frac{100!}{33!33!34!} = 4.1924 \times 10^{45}, \tag{4.4}$$

which is a very large number. Indeed it is so large that most current computers would be unable to compute this many permutations of the data in an acceptably short time span. Consequently, when doing a permutation test of the data, an investigator would need to compute some number N of permutations that is much less than M. Most programming languages contain programs to compute random numbers, which can be used to write code for computing a sample of random permutations of a data set.

When one is doing a permutation test, the null hypothesis H_0 is that the observed data set is a random sample from the set $\mathbb{S}$ of all possible permutations of the data. To carry out such a test, a computer would need to be programmed in such a way that some number N of random permutations of the data would be computed and for each permutation of the data estimates of the effects of the vector $\mathbf{E}$ in equation (3.26) of Section 3 would be computed. Let **SIM** denote a $5 \times N$ matrix of simulated estimates of the five effects in the vector $\mathbf{E}$ such that each column of this matrix is an estimated realization of the observed vector $\mathbf{E}$ based on a random permutation of the data. Similarly, let **ALPHA** denote a $5 \times N$ matrix such that each column is a copy of the vector $\mathbf{E}$. Then, consider the relationship $\geq$ and a $5 \times N$ matrix $\mathbf{R}$ defined by

$$\mathbf{R} = \mathbf{SIM} \geq \mathbf{ALPHA}. \tag{4.5}$$

Let $\mathbf{SIM}(i,j)$ denote the element from the ith row and jth column of the matrix **SIM**, and define the element $\mathbf{ALPHA}(i,j)$ analogously. Each element of the matrix $\mathbf{R}$ is 0 or 1. If the relation

$$\mathbf{SIM}(i,j) \geq \mathbf{ALPHA}(i,j) \tag{4.6}$$

is true, then the element $\mathbf{R}(i,j) = 1$, and if this relation is false, $\mathbf{R}(i,j) = 0$. As will be demonstrated in what follows, by using the matrix $\mathbf{R}$ various types of p values may used to judge the statistical significance of each of the estimated effects.

For example, let the SUMROWS denote the array with five elements that results from summing the rows of the matrix $\mathbf{R}$, and let $r(\nu)$ denote the νth element in this array, where $\nu = 1, 2, \ldots, 5$. For example, according to the ordering used in defining the 5×1 vector $\mathbf{E}$ in Section 3, the number $r(\nu)$ denotes the number of times among the N sample values that the inequality

$$\mathbf{SIM}(1,j) \geq \mathbf{ALPHA}(1,j) \tag{4.7}$$

was satisfied for effect 1.

Observe that for every $\nu = 1, 2, \ldots, 5$ the value of $r(\nu)$ will be a member of the set $\{x | x = 0, 1, 2, \ldots, N\}$. Consequently, $p(\nu) = r(\nu)/N$ is the estimated p value for the estimated effect $\alpha^2(\nu)$, for $\nu = 1, 2, \ldots, 5$; see the vector $\mathbf{E}$ in (3.26) in Section 3 for details. Note that according to the elements in this vector, $p(1)$ is the p value for the estimated effect $\alpha^2(1)$ of the marker allele 1. If for any $\nu = 1, 2, \ldots, 5$, $p(\nu) < 0.05$, then the null hypothesis H_0 that the data are a sample from the set $\mathbb{S}$ of all permutations of the data will be rejected and the estimate of the effect $\alpha(\nu)$ will be said to be statistically significant. In this example, the probability 0.05 was chosen arbitrarily, but an investigator would be free to choose any other small probability as a benchmark for declaring statistical significance. Alternatively, the investigator may want to observe each p value, and make a judgment as to whether an estimated effect was statistically significant. At this point in the discussion, note that each of the probabilities estimated from the data pertains to only the statistical significance of one of the five estimated effects under consideration. But, as will be demonstrated below, it will also be possible to obtain joint probabilities that some sets of effects are jointly significant, which will be illustrated in the next paragraph.

This other set of interesting joint p values may be computed by summing the columns of the matrix $\boldsymbol{R}$. Let SUMCOLUMNS denote an array with N elements that are sums of the columns of $\boldsymbol{R}$, and let $c(\nu)$ denote the sum of column ν. Then, the value of each $c(\nu)$ is an integer in the set $\{y | y = 0, 1, 2, 3, 4, 5\}$ for $\nu = 1, 2, \ldots, N$. For example, if for some column ν, $c(\nu) = 0$, then all the numbers in column ν of $\boldsymbol{R}$ would be 0, indicating that for every row i the inequality (4.6) was not satisfied for column $\nu = j$. But, if $c(\nu) = 5$ for some column $\nu = j$, then the inequality in (4.6) would be satisfied for all rows $i = 1, 2, 3, 4, 5$. Given the array SUMCOLUMNS, it would be possible to estimate a distribution that would provide insights into the joint statistical significance of the five estimated effects in the column E in (3.26) in Section 3. For any fixed $\nu = 0, 1, 2, 3, 4, 5$, let $m(\nu)$ be the number of elements of the array SUMCOLUMNS has the value ν. Let $p_{\text{joint}}(\nu) = m(\nu)/N$ denote the estimated probability for the values $\nu = 0, 1, 2, 3, 4, 5$. By viewing these joint probabilities, an investigator would be in a position to judge whether all the five estimated effects were jointly statistically significant. If, for example, this distribution were skewed to the left so that the probabilities $p_{\text{joint}}(\nu)$ for $\nu = 0, 1$ were larger than the probabilities $p_{\text{joint}}(\nu)$ for $\nu = 4, 5$, then the investigator could make a judgment as to whether all estimates of the five effects were jointly statistically significant.

Some investigators may wish to carry out a permutation test, but in this chapter the focus of attention will be on another class of tests of statistical significance based on Monte Carlo simulation methods that will be formulated in the next section. However, in practice, an investigator may want to carry out statistical tests of significance belonging to different classes of tests to get some idea as to whether estimated effects are statistically significant for at least two classes of tests of statistical significance.

5. Testing the Statistical Significance of Estimated Effects Based on Monte Carlo Simulation Methods

There is another approach to judging whether estimates of the five effects are statistically significant by using Monte Carlo simulation methods. Suppose, for example, that there are $n(i,j)$ nonnegative simulated realizations of the random variable $W(i,j)$ for every genotype $(i,j) \in \mathbb{G}_2$. The rationale underlying the choice of nonnegative random variable is that most measures with respect to some quantitative trait are nonnegative numbers. One of the simplest approaches to simulation realizations of nonnegative random variables is to use the absolute or folded normal distribution. Suppose that a random $X(i,j)$ has a normal distribution with an expectation $\mu(i,j)$ and variance $\sigma^2(i,j)$ for every genotype $(i,j) \in \mathbb{G}_2$. Let Z denote a standard normal random variable with expectation 0 and variance 1. Then, as is well known, if Z is a simulated realization from a standard normal distribution, $X(i,j) = \mu(i,j) + \sigma(i,j)Z$ is a simulated realization of the random variable $X(i,j)$ for every genotype $(i,j) \in \mathbb{G}_2$. Given a simulated realization of a random variable $X(i,j)$, $W(i,j) = |X(i,j)|$ is a simulated realization of a random variable $W(i,j)$ with an absolute normal distribution for every genotype $(i,j) \in \mathbb{G}_2$. A more detailed description of the folded normal distribution will be given in an appendix.

The next step in setting up a Monte Carlo simulation experiment is to formulate a procedure for testing a null hypothesis. Suppose, for example, that the $n(i,j)$ observations of the random variable $W(i,j)$ for every genotype $(i,j) \in \mathbb{G}_2$ are a random sample from a folded normal distribution. Moreover, suppose that there are real data consisting of a sample of size $n(i,j)$ for each genotype $(i,j) \in \mathbb{G}_2$. To simplify the notation, from now on the symbols $\mu(i,j)$ and $\sigma^2(i,j)$ will denote estimates of the corresponding expectation and variance for each genotype $(i,j) \in \mathbb{G}_2$ based on the data. Given the data, an investigator could also estimate the genotypic distribution $\{p(i,j)|(i,j) \in$

$\mathbb{G}_2\}$ as well as the 5×1 vector E of effects. In this situation, the investigator may wish to entertain the null hypothesis H_0 of homogeneity and suppose that there are positive numbers μ and σ^2 such that

$$\mu(i,j) = \mu_{uc},\ \sigma^2(i,j) = \sigma^2_{nc} \tag{5.1}$$

for all genotypes $(i,j) \in \mathbb{G}_2$, the subscript uc stands for "unconditional."

At this point, to simulate samples from a normal distribution with expectation μ_{uc} and variance σ^2_{uc}, an investigator may decide to choose μ_{uc} and σ^2_{uc} as

$$\mu_{uc} = \sum_{(i,j)\in\mathbb{G}_2} p(i,j)\mu(i,j), \tag{5.2}$$

$$\sigma^2_{uc} = \sum_{(i,j)\in\mathbb{G}_2} p(i,j)\sigma^2(i,j), \tag{5.3}$$

as the parameters in the normal distribution to be used to simulate $n(i,j)$ realizations of the random variable $X(i,j)$ for every genotype $(i,j) \in \mathbb{G}_2$. Given this simulated sample of realizations of random variables with a homogeneous distribution, the simulated $n(i,j)$ realizations of the phenotypic random variables $W(i,j)$ would be computed by using the formula $W(i,j) = |X(i,j)|$ for every genotype $(i,i) \in \mathbb{G}_2$.

This choice of σ^2_{uc}, for example, would result in a simulated sample with a variance that would be close to that in the original data so that unrealistic outliers would occur with small probabilities in the simulated data. The choice of μ_{uc}, however, is less sensitive than that for σ^2_{uc}. For it is interesting to note that if the hypothesis H_0 were true, then all effects in the vector E would be 0 for any choice of μ. For example, consider the first order effect

$$\alpha(1) = \mu(1) - \mu; \tag{5.4}$$

see (3.14) in Section 3. By inspecting (3.13) in Section 3, it can be seen that if H_0 is true, then $\mu(1) = \mu$ so that $\alpha(1) = 0$.

Similar arguments may be used to see that if H_0 is true, then all the effects in the vector E will be 0. It is interesting to note that for any choice of μ in a simulation procedure, all the effects in the vector E would be 0. From the point of view of simulating $n(i,j)$ realizations of the random variables $W(i,j) = |X(i,j)|$ for every genotype $(i,i) \in \mathbb{G}_2$, it can be seen that if the expectations of the random variables $X(i,j)$ are some constant μ, then the expectations of the random variables $W(i,j)$ would also be some constant. Hence, the estimated effects based on the means of folded normal distribution would also be constant. If the reader is interested in further details regarding the folded normal distribution, it is suggested that the appendix be consulted.

To carry out a test of statistical significance using the Monte Carlo simulation procedure just outlined, an investigator would start with the array **DATA** in (2.2). Then, the first step in a Monte Carlo simulation procedure would be that of simulating a quasi-data set, **QUASIDATA**, consisting of n realizations of a random W with a normal distribution with expectation μ_{uc} and variance σ^2_{uc}, as indicated above. Let $\mathbf{SIM}W(1,1)$ denote the first $n(1,1)$ elements of the array, **QUASIDATA**, and define the arrays $\mathbf{SIM}W(0,0)$ and $\mathbf{SIM}W(1,0)$ similarly for the numbers of individuals $n(0,0)$ and $n(1,0)$ of genotypes $(0,0)$ and $(1,0)$, respectively. Then, the simulated quasi-array of data for the three genotypes may be represented in the form

$$\mathbf{QUASIDATA} = \mathbf{SIM}W(1,1), \mathbf{SIM}W(0,0), \mathbf{SIM}W(1,0), \tag{5.5}$$

just as the real data in (4.2).

Given the simulated data set in (5.5), the next step in the Monte Carlo simulation procedure would be that of computing estimates of the five effects in the 5×1 vector $\boldsymbol{E}$. By repeating this step just outlined $N > 1$ times, a version of the $5 \times N$ matrix $\boldsymbol{R}$ in (4.5) could be computed. Then, by using the procedure outlined in Section 4, a test of statistical significance for any estimated effect in the vector $\boldsymbol{E}$ could be accomplished. Similarly, a test for the joint statistical significance of the five estimated effects could be carried out, by using the procedure outlined in Section 4 by summing the rows of the matrix $\boldsymbol{R}$. In such an experiment, an investigator would be testing the null hypothesis that all the five effects in the vector $\boldsymbol{E}$ are zero.

To simplify the notation, from now on the symbols $\mu(i,j)$ and $\sigma^2(i,j)$ will denote estimates of the corresponding expectation and variance for each genotype $(i,j) \in \mathbb{G}_2$. The rationale for considering simulated nonnegative random variables is that the majority of measurements for some quantitative traits are usually nonnegative numbers, as stated above. Given this information, an investigator could estimate the genotypic distribution $\{p(i,j) | (i,j) \in \mathbb{G}_2\}$ as well as the 5×1 vector $\boldsymbol{E}$ of effects. In setting up this Monte Carlo experiment, the investigator may wish to entertain the null hypothesis H_0 that there are positive numbers μ and σ^2 such that

$$\mu(i,j) = \mu_{uc},\ \sigma^2(i,j) = \sigma^2_{uc} \tag{5.6}$$

for all genotypes $(i,j) \in \mathbb{G}_2$; see (5.2) and (5.3).

One of the principal goals of the type of Monte Carlo simulation under consideration is that of computing p values on which a judgment of statistical significance for each of the five effects that were estimated from the data can be made. The $5 \times N$ matrix $\boldsymbol{R}$ plays a fundamental role in estimating the p values. Consider, for example, an element-by-element representation of this matrix of the form

$$\boldsymbol{R} = [r_{ij}] \tag{5.7}$$

and let $\alpha^2(i)$ denote the squared effect estimated from the data in the vector $\boldsymbol{E}$ for $i = 1, 2, 3, 4, 5$. Similarly, let $\alpha^2(i,j)$ be an estimate of the squared effect i of replication j of the Monte Carlo simulation experiment for $j = 1, 2, \ldots, N$. Then, for each $i = 1, 2 \ldots, 5$ and $j = 1, 2, \ldots, N$, r_{ij} may be interpreted as a Bernoulli indicator such that $r_{ij} = 1$ if the event

$$[\alpha^2(i,j) \geq \alpha^2(i)] \tag{5.8}$$

occurs and $r_{ij} = 0$ if the event defined in (5.8) does not occur.

Given a null hypothesis H_0, let

$$E[r_{ij}|H_0] = 1P[r_{ij} = 1|H_0] + 0P[r_{ij} = 0|H_0] = P[r_{ij} = 1|H_0] = p(i|H_0) \tag{5.9}$$

denote the conditional probability that the event in (5.8) occurs, given H_0 for $j = 1, 2, \ldots, N$. The technical details of the random number generator used in the Monte Carlo simulation experiments reported in the following sections will not be discussed here. But, because the randomness in the properties of the sequences of uniform random numbers taking values in the interval $[0, 1)$, the assumption that the Bernoulli indicators denoted in (5.10) are independent for all $j = 1, 2, \ldots, N$ so that for each $i = 1, 2, \ldots, 5$ it is highly plausible to assume that the sequence of Bernoulli indicator functions are independently distributed with a common expectation $p(i|H_0)$ defined in (5.11). Therefore, by invoking the law of large numbers, it follows that

$$\lim_{N\uparrow\infty} \frac{1}{N} \sum_{j-1}^{N} r_{ij} = p(i|H_0), \tag{5.10}$$

for every $i = 1, 2, \ldots, 5$. If the strong law of large numbers is invoked, then the limit in (5.10) holds with probability 1. In general, the larger the choice of the number N, the greater the reliability of the estimate in (5.12). In the experiments that will be reported in subsequent sections of this chapter, N was chosen as 10,000. With this choice of N the computing run time of each of the experiments reported in the sections to follow was in the range of 2 to 3 minutes, which would be acceptable if a quantitative trait under consideration involved repeating an experiments for 10 to 15 loci which were thought to be involved in the expression of the trait.

At this point in the development of ideas making up the procedure for tests of significance under consideration, it will be helpful to express the ideas in the last paragraph of Section 4 more formally. Let

$$s_j = \sum_{i=1}^{5} r_{ij} \tag{5.11}$$

denote the sum of the indicators in column j or the matrix R for $j = 1, 2, \ldots, N$. Then, the array SUMCOLUMNS defined in the last paragraph of Section 4 has the form

$$\text{SUMCOLUMNS} = (s_1, s_2, \ldots, s_N). \tag{5.12}$$

Let

$$[j|s_j = x], \tag{5.13}$$

for $x = 0, 1, 2, 3, 4, 5$. Then,

$$m(x) = \sum_{j\in[j|s_j=x]} s_j \tag{5.14}$$

for all x and

$$\sum_{x=0}^{5} m(x) = N. \tag{5.15}$$

Then, in terms of the formal system developed in this section,

$$p_{\text{joint}}[x|H_0] = \frac{m(x)}{N} \tag{5.16}$$

for all x is the conditional distribution for judging the joint statistical significance of the five effects under consideration. Observe that the conditional distribution defined in (5.16) has the property

$$\sum_{x=0}^{5} p_{\text{joint}}[x|H_0] = 1, \tag{5.17}$$

as it should. In any simulation experiment, it is useful to check that this equation holds as part of tests for the correctness of the software. Equation (5.19) is justified, because sets in the collection of sets in (5.15) are a disjoint partition of the set SUMCOLUMNS in (5.14).

6. Monte Carlo Simulation Experiments on Simulating Data and Testing the Null Hypotheses

There is a vast literature on Monte Carlo simulation procedures that have been used in many fields of science. For example, the paper by Mode and Gallop (2008) [1], as it turned out, provided the authors with a window on an extensive literature on Monte Carlo simulation procedures as used in many fields of science; see the Internet link http://biomedupdater.com/urlu8c?srk=2a20a8d6419d877ad74198186ff9a1c0ae53b88acaa0ade5587ca0baf302b72.

Furthermore, the book cited in [2] and edited by the author also contains an extensive collection of papers on the application of Monte Carlo simulation methods in various fields of biology and related sciences. In this section, the random number generator with a very long period set forth in Mode and Gallop (2008) [1] in Section 2 on Monte Carlo simulation methods will be used as well as in all sections to follow, containing accounts of Monte Carlo simulation experiments used to test various versions of the null hypothesis. There is a caveat that the reader should be aware of when reading the experimental results reported in this and the following sections; namely, the random number generator used in all experiments reported in this paper was designed for computers based on 32-bit words. The computers used to conduct Monte Carlo simulation experiments reported in this paper, however, were based on 64-bit words. Algorithms for random number generators for 64-bit words may be found in the papers cited in Mode and Gallop (2008) [1], but to implement these algorithms for use on computers based on 64-bit words would require an extensive period of development, using array-manipulating programming languages such as APL. It seems plausible, however, that if the Monte Carlo simulation experiments reported in this paper were based on a random number generator designed for a 64-bit word computer rather than the 32-bit word generator, the results and conclusions would not be significantly different.

The first step in setting up a Monte Carlo experiment to test some null hypothesis H_0 is to simulate the data that will be used to estimate all parameters and effects as functions of parameters. In the best of all worlds, data of the type under consideration would be posted on the Internet so that it could be downloaded by investigators and used to present concrete examples of the application of new statistical procedures. But, unfortunately, getting permission to use such data is often impossible, unless you are a member of the group that has assembled the data. The parameter values used to simulate the data used in the experiments discussed in the section are shown in Table 6.1.

By way of interpreting the chosen values in Table 6.1, on some scale of hypothetical units used to measure the expression of some quantitative trait under consideration, the expected values for the three genotypes $(1,1)$, $(0,0)$ and $(1,0)$ were chosen as $\mu(1,1) = 30$, $\mu(0,0) = 40$ and $\mu(1,0) = 60$. The rationale used in choosing these numbers was to assign different values to each of these expected values so that the estimated genetic variance would be positive. The rational for not choosing $\mu(1,0) = 50$ but $\mu(1,0) = 60$

Table 6.1. Parameter Values Used to Simulate Data

$\mu(1,1)$	30
$\sigma(1,1)$	$0.25\mu(1,1)$
$\mu(0,0)$	40
$\sigma(0,0)$	$0.25\mu(0,0)$
$\mu(1,0)$	60
$\sigma(1,0)$	$0.25\mu(1,0)$

was to consider the case where there was a heterotic effect for heterozygotes of genotype $(1, 0)$, i.e. it was assumed that there was some interaction of alleles 1 and 0 in individuals, whose genotype was the heterozygote. The reason for choosing the standard deviations $\sigma(1, 1)$, $\sigma(0, 0)$ and $\sigma(1, 0)$ as a common fraction 0.25 of the expectation for each genotype was that the estimates of the environmental valance for each genotype seemed plausible, as observed in preliminary experiments. The sample sizes chosen for the genotypes were $n(1, 1) = 100$, $n(0, 0) = 200$ and $n(1, 0) = 450$. These sample sizes resulted in the allele frequencies $p(1) = 0.433$ and $p(0) = 0.567$. The estimated heritability based on the simulated data for the three genotypes was $H = 0.4280$.

Two null hypotheses were considered in the illustrative examples on testing the statistical significance of the squared effects estimated from the simulated data. Presented in Table 6.2 are the assigned parameter values used in testing the two null hypotheses under consideration.

In Table 6.2, the subscript uc stands for unconditional expectations and standard deviations as shown in (5.1). Moreover, the estimates of these parameters were computed from the simulated data, using formulas (5.2) and (5.3). The estimate σ_{uc} was computed using the formula

$$\sigma_{uc} = \left(\sigma_{un}^2\right)^{\frac{1}{2}}; \tag{6.1}$$

see (5.2). Table 6.3 contains the symbolic form of the squares of the estimated effects, the estimates of these parameters based on the simulated data and the p values computed in tests of statistical significance of the null hypotheses $H_0(1)$ and $H_0(2)$ using the Monte Carlo simulation methods under consideration.

Table 6.2. Parameter Values Used in Testing Two Null Hypotheses Based on Monte Carlo Simulation Methods

Parn	$H_0(1)$	$H_0(2)$
$\mu(1, 1)$	μ_{uc}	0
$\sigma(1, 1)$	σ_{uc}	σ_{uc}
$\mu(0, 0)$	μ_{uc}	0
$\sigma(0, 0)$	σ_{uc}	σ_{uc}
$\mu(1, 0)$	μ_{uc}	0
$\sigma(1, 0)$	σ_{uc}	σ_{uc}

Table 6.3. Statistical Test of Significance of the Estimates of the Squared Effects Based on Monte Carlo Methods

ESQ	EST	$H_0(1)$	$H_0(2)$
$\alpha^2(1)$	0.03185	0.6124	0.3948
$\alpha^2(0)$	0.00557	0.7796	0.64411
$\alpha^2(1,1)$	376.7118	0	0
$\alpha^2(0,0)$	92.48211	0	0
$\alpha^2(1,0)$	106.8706	0	0

In the simulation experiment designed to test the null hypothesis $H_0(1)$, the squares of the estimated effects were computed with 10,000 Monte Carlo replications, using the parameter assignments listed in the second column of Table 6.2. Similarly, to test the null hypothesis $H_0(2)$, 10,000 Monte Carlo replications of the squared effects were computed. The p values listed in columns 3 and 4 of Table 6.3 were computed using the 10,000 Monte Carlo replication as set forth in equation (5.12), with $N = 10{,}000$ for each null hypothesis being tested. From rows 1 and 2 of Table 6.2, it can be seen that if the null hypothesis $H_0(1)$ is true, then $\alpha^2(1) = 0$ and $\alpha^2(0) = 0$. The estimates of these two squared effects based on the simulated data are 0.03185 and 0.00557, with the corresponding p values 0.6124 and 0.7796, respectively, under the null hypothesis $H_0(1)$, and are not sufficiently small for one to reject the null hypothesis being tested. An investigator may therefore conclude that the additive effects of alleles 1 and 0 are not statistically different from 0. As can be seen from the second column of Table 6.3, the estimates of the squared effects for interactions effects $\alpha^2(1,1)$, $\alpha^2(0,0)$ and $\alpha^2(1,0)$ are 376.7118, 92.48211 and 106.8706, respectively, with the corresponding p values 0, 0, 0. The number 0 is the smallest possible p value; consequently, the squared effects for allelic interaction are highly significantly different from zero under the null hypothesis $H_0(1)$. It is interesting to note that if the null hypothesis $H_0(2)$ were tested using the same methods, the statistical conclusions just stated for the five squared effects under consideration would not change even though the p values for the additive effects differ from those that were computed under the null hypothesis $H_0(1)$.

Table 6.4. Estimates of the Joint *p*-Values Under Each Null Hypothesis

Values	0	1	2	3	4	5
$H_0(1)$	0.2204	0.1672	0.6124	0	0	0
$H_0(2)$	0.3559	0.2403	0.3948	0	0	0

The last set of p values that will be presented in this section are those described in (5.13) through (5.21), which were estimated by summing the columns $5 \times 10{,}000$ matrix R of realized Bernoulli indicator functions as described in Section 5. Presented in Table 6.4 are the are estimates of these p values under null hypotheses $H_0(1)$ and $H_0(2)$.

Observe that for each row in this table the listed p values are a distribution satisfying equation (5.21). For example, for the null hypothesis $H_0(1)$,

$$0.2204 + 0.1672 + 0.6124 = 1. \tag{6.2}$$

An equation of the same form would also be valid for the row in Table 6.4 corresponding to the null hypothesis $H_0(2)$. By way of interpreting this table, with an estimated probability 0.2204, the sum of a column of the indicator matrix R would be zero under the null hypothesis $H_0(1)$. Similarly, under this null hypothesis, 0.6124 was the estimated probability that the sum of a column of indicators was 2. Observe that, because this probability is largest in the distribution under the null hypothesis $H_0(1)$, the number 2 is the mode of this distribution of joint p values. It is also interesting to note that the estimated probability that a column sum of indicators in the matrix R had the value $3, 4$ or 5 was 0 under both null hypotheses under consideration. It is interesting to also note that the number 2 was also the mode of the distribution of joint probabilities under the null hypothesis $H_0(2)$. Moreover, under both null hypotheses, the events that two columns of the indicator matrix R both occurred more often and has a greater influence on the p values for judging the statistical significance of each squared effect that was estimated by summing the rows of the indicator matrix R.

Software to test the statistical significance of the estimated heritability $H = 0.4280$ was also developed so that p values of testing null hypotheses of the form H_0: $H = 0$ could be tested in Monte Carlo simulation experiments. For all the null hypothesis tests described in this section, the p value was zero. Hence, the estimate of heritability was, in a statistical sense, highly significantly different from zero.

7. Two Monte Carlo Simulation Experiments to Simulate Data and Test Null Hypotheses

In this section two sets of simulated data, for experiments A and B, will be used in testing null hypotheses. Presented in Table 7.1 are the chosen parameter values for simulation of data used in experiments A and B reported in this section.

From this table it can see seen that, with the exception of the sample sizes $n(1,1), n(0,0)$ and $n(1,0)$, the values of the parameters chosen for experiment A in the second column of Table 7.1 are the same as those in Table 6.1. In experiment A, it was assumed that allele 1 was a rare mutation that arose in some ancestral population from which the sample evolved. The number of individuals of genotype $(1,1)$ in the sample was chosen as $n(1,1) = 10$, and the sample sizes for genotypes $(0,0)$ and $(1,0)$ were chosen as $n(0,0) = 1{,}000$ and $n(1,0) = 15$. The objective of experiment A was to provide some insights concerning what impact the low frequency of allele 1 in the sample would have on the estimates and test of statistical significance reported in Section 6. With the exception of the standard deviations, which were chosen as $\sigma(i,j) = \mu(i,j)$ for all genotypes $(i,j) \in \mathbb{G}_2$, the sample sizes and expectations of a quantitative trait under consideration were the same

Table 7.1. Parameter Values Used for Simulation of Data in Experiments A and B

Prams	Exp A	Exp B
$\mu(1,1)$	30	30
$\sigma(1,1)$	$0.25\mu(1,1)$	$\mu(1,1)$
$\mu(0,0)$	40	40
$\sigma(0,0)$	$0.25\mu(0,0)$	$\mu(0,0)$
$\mu(1,0)$	60	60
$\sigma(1,0)$	$0.25\mu(1,0)$	$\mu(1,0)$
$n(1,1)$	10	100
$n(0,0)$	1,000	200
$n(1,0)$	15	450

Table 7.2. Parameter Values for Testing Null Hypotheses

Parms	Exp A	Exp B
$\mu(1,1)$	$\mu_{uc}(\text{A})$	$\mu_{uc}(\text{B})$
$\sigma(1,1)$	$\sigma_{uc}(\text{A})$	$\sigma_{uc}(\text{B})$
$\mu(0,0)$	$\mu_{uc}(\text{A})$	$\mu_{uc}(\text{B})$
$\sigma(0,0)$	$\sigma_{uc}(\text{A})$	$\sigma_{uc}(\text{B})$
$\mu(1,0)$	$\mu_{uc}(\text{A})$	$\mu_{uc}(\text{B})$
$\sigma(1,0)$	$\sigma_{uc}(\text{A})$	$\sigma_{uc}(\text{B})$

as those in Table 6.1. The objective of experiment B was to provide some insights into the effects which higher environmental variances would have on the estimates of parameters when compared with the experiments and tests of statistical significance reported in Section 6.

In experiment A, the estimates of the frequencies of alleles 1 and 0 were about $p(1) = 0.01$ and $p(0) = 0.9$. The estimate of heritability in experiment A was $H_{\text{A}} = 0.0902$. In experiment B, the estimates of the frequencies of alleles 1 and 0 were $p(1) = 0.4333$ and $p(0) = 0.5667$, and the estimate of heritability was $H_{\text{B}} = 0.0932$.

Contained in Table 7.2 are the parameter values used to test the null hypotheses in experiments A and B.

In Table 7.2 the same subscripts are used for the designated parameter values used to test to define the null distributions for testing null hypotheses in experiments A and B. The values of these parameters were chosen by using the formulas in equations (5.2) and (5.3), but it is clear from the assigned parameter values in Table 7.1 that the estimated values of the parameters in Table 2 would differ in experiments A and B.

Table 7.3 contains estimates of the squares of effects and p values estimated by using 10,000 Monte Carlo replications in both experiments A and B.

In Table 7.3, the second column contains the estimates of the squared effect using the simulated data. With the exception of the estimate of the additive effect $\alpha^2(0)$, which was 99.4566, the estimated squares of the remaining effects have zero p values, and are highly statistically different from zero. By way of contrast, in Table 6.3 all the squared additive effects are not statistically different under either null hypotheses $H_0(1)$ or $H_0(2)$. From

Table 7.3. Estimated Squared Effects and p Values for Experiments A and B

Parms	Est A	p values A	Est B	p values B
$\alpha^2(1)$	101.5387	0	54.4873	0
$\alpha^2(0)$	99.4566	0.1525	56.0312	0
$\alpha^2(1,1)$	105.7899	0	1771.4965	0
$\alpha^2(0,0)$	403.1152	0	1043.0865	0
$\alpha^2(1,0)$	409.493	0	7.1615	0.0111

Table 7.4. Estimates of the Joint p Values Under Each Null Hypothesis

Values	0	1	2	3	4	5
$H_0(A)$	0.8475	0.1525	0	0	0	0
$H_0(B)$	0.9889	0.0111	0	0	0	0

this example, it can be seen that the small numbers of genotypes $(1,1)$ and $(1,0)$ had a significant effect on the reported p values in column 3 of the table. In column 5, where the p values for experiment B are displayed, it can be seen from the estimated p values that, with the exception of the estimate of the squared effect $\alpha^2(1,0)$, the estimates of the other four squared effects are highly statistically different from zero. It is also interesting to note that from an estimated p value of 0.0111, it may be concluded at about the one-percent level, the estimate 7.1615 of $\alpha^2(1,0)$ is statistically different from zero. The results of this experiment suggest that, even with a low estimate of heritability resulting from higher assigned values of the environmental variances, there may be interesting cases using real data that would lead to squared effects that were statistically different from zero.

The next set of p values to be presented in this section are two joint distributions for each of the null hypotheses under consideration, as shown in Table 7.4.

From Table 7.4, it can be seen that for both hypotheses the mode of the distribution was 0. It is also interesting to note that for the hypothesis $H_0(A)$ the p value at the number 1 is the same as the p value in Table 7.3 corresponding to the estimate of the parameter $\alpha^2(0)$. Similarly, the p value for the hypothesis $H_0(B)$ at the number 1 is the same as the p value corresponding to the estimate of the parameter $\alpha^2(1,0)$ in Table 7.3. From these observations, it follows that for both null hypotheses there were no columns of the $5 \times 10{,}000$ matrix R with ones for the values 2, 3, 4, 5. It should also be mentioned that the estimates of heritability for both hypotheses had zero p values even though both estimates were small. Recall that $H_{\mathrm{A}} = 0.0902$ and $H_{\mathrm{B}} = 0.0932$.

8. Further Developments and Potential Applications

By using various statistical and other methods, researchers have identified a number of regions in the human genome that are associated with diseases such as Alzheimer's. A recent paper on such regions has been reported in Raj *et al.* (2012) [3], in which, among other things, 11 regions of the human genome, associated with susceptibility to Alzheimer's disease, have been identified. Evidence is also reported on the existence of a protein network involving four of these regions that is sustained in the human genome by natural selection. The number of individuals in this sample is about 5,000 times that the DNA of each individual in the sample has been sequenced. In a related paper Rossin *et al.* (2011) [4] reported that proteins coded by identified regions of the human genome associated with immune-mediated diseases physically interact and suggest some underlying basic biology. Alzheimer's disease may also be viewed as a quantitative trait whenever its expression is measured on some numerical scale. Moreover, if each of the 11 genomic regions may be identified in two alternative forms, then from the point of view of quantitative genetics these 11 regions may be referred to as loci with two alleles at each locus.

When an investigator considers 11 loci and two alleles per locus, the number of effects that may be estimated directly will become very large even if only three genotypes per locus may be identified, as discussed in the published Chapters 3 and 4. For example, for the case of four loci for which only three genotypes may be identified per locus, the number of identifiable genotypes with respect to 11 loci would be

$$3^{11} = 1.7715 \times 10^{5}. \tag{8.1}$$

However, if only four loci were under consideration, then the number of identifiable genotypes would be

$$3^{4} = 81. \tag{8.2}$$

As expected, this is a much smaller number than that in (8.1), but, nevertheless, when an array with 81 cells is under consideration, a problem that may arise is whether the number of individuals in each cell is large enough for one to draw statistically reliable statistical inferences. Such problems suggest that an investigator should explore the data to estimate the genotypic frequency of each genotype as well the frequencies of each allele at the four loci under consideration. Similar questions will arise whenever an investigator wishes to explore the data to determine whether there are a sufficient number of observations in each cell for one to draw reliable statistical inferences, when N, the number of loci under consideration, is such that $N > 4$. A step that should be included in any exploration of the data would be that of determining if all loci under consideration were autosomal, for if one or more sex-linked loci are included in the sample, then such loci would need to be treated separately.

When all the loci are autosomal, one approach to determining the number of loci that are such that each cell in a multidimensional array would have a sufficiently large number of observations for one to draw reliable statistical inferences is to investigate each locus under consideration. In this investigation one of the goals would be to determine, among other things, whether the frequency of the two alleles at each locus is sufficiently large enough to be included in the construction of arrays of data with respect to two or more loci that will contain a sufficient number of observations in each cell for one to draw reliable statistical inferences. For the case of the data on Alzheimer's disease mentioned above, an investigator would need to do an exploratory experiment involving 11 loci with two alleles at each locus. But, even in a sample of 5,000 individuals, the frequency of some alleles at one locus or two or more loci may not be sufficiently large for one to construct multidimensional arrays that involve low frequency alleles.

These observations suggest that it would be expedient for the above case of a sample of 5,000 individuals to estimate the frequency at each of the 11 loci to obtain information as to whether each allele at each locus has a sufficiently high frequency to be included in multidimensional arrays with respect to two or more loci. But there are other criteria that could be used to judge as to what loci would be included in multidimensional arrays. For example, if an estimate of heritability at some locus is low, then including this locus in a multidimensional array may not be fruitful. An investigator could also use estimates of the five effects that may be estimated when one autosomal locus is under consideration and carry out a statistical test of significance on the squares of the effects to judge which ones are statistically significant for each of the 11 loci. If there were loci for which none of the squares of effects were not statistically significant, then the investigator might not want to include this locus in a multidimensional array involving two or more loci.

With regard to further developments of the software, it would be possible to create a front end to the APL programs to do the analyses reported in this paper so that the existing APL software could be used to carry out the type of exploratory experiment described above using any computer platform. But, before data consisting of multiple arrays involving two or more autosomal loci can be analyzed, an investigator would need to either find existing software or write software to accommodate multidimensional arrays of data on two or more autosomal loci. If APL were used to write this software, then the existing software for the case of two alleles at one autosomal locus could become part of the extended software when the additive effects and intralocus effects are estimated at each of the loci under consideration. But, to estimate effects involving two or more loci, new programs would need to be written. It seems very plausible that an array-manipulating programming language such as APL would be helpful in writing succinct code designed to process multidimensional data on multiple loci. If an investigator wanted to consider two or more quantitative traits, then the ideas in Chapter 4 could be modified to accommodate the case that only three genotypes at each locus could be recognized. Moreover, the APL software used in this chapter could be extended to take into account two or more traits at each locus. For those cases in which two or more loci and two or more traits are under consideration, an array-manipulating programming such as APL would be very helpful in writing code to do the required matrix operation described in Chapter 4. Just like the ordering of the three genotypes considered in this chapter which played a basic role in writing the software, some expeditious ordering of the genotypes with respect to two or more loci will also be a crucial step in developing computer code to accommodate cases of multiple loci with three recognizable genotypes at each locus.

APPENDIX

As an aid to developing a deeper understanding as to the properties of the absolute normal distribution that was used in Monte Carlo simulation experiments described in previous sections, in this appendix formulas for the expectation and variance of this distribution will be derived. In the literature on probability and statistics, the absolute normal distribution is called the folded normal, and for the case $\mu = 0$ and $\sigma = 1$, this distribution is known as the half normal. The formulas that will be derived below may be found on the Internet, and if the reader is interested in more details, it is suggested that the website en.wikipedia.org/wiki/Folded_normal_distribution be consulted, where some references are also listed. Some proofs of the formulas derived below may also be found on the Internet, but many of these proofs lack transparency. In what follows, attempts will be made to include enough details with the hope that the derivation of the formulas will be transparent.

The first distribution to be described is the half normal. Let Z denote a normal random variable with expectation 0 and variance 1. In symbols, $Z \sim N(0,1)$. The *pdf* of Z is

$$\varphi(z) = \frac{1}{\sqrt{2\pi}} \exp\left[-\frac{1}{2}z^2\right] \tag{A.1}$$

for $z \in (-\infty, \infty) = \mathbb{R}$, the set of real numbers. The distribution function of Z is, therefore,

$$\Phi(z) = \frac{1}{\sqrt{2\pi}} \int_{-\infty}^{z} \exp\left[-\frac{1}{2}y^2 dy\right] \tag{A.2}$$

for $z \in \mathbb{R}$. Let $\mu \in \mathbb{R}$ and $\sigma \in [0, \infty)$. Then, by definition, a random variable $X = \mu + \sigma Z$ has a normal distribution with expectation

$$E[X] = \mu + \sigma E[Z] = \mu \tag{A.3}$$

and variance

$$\text{var}[X] = E\left[(X-\mu)^2\right] = \sigma^2. \tag{A.4}$$

The random variable $Y = |Z|$, the absolute value of Z, maps $\mathbb{R}$ into $[0, \infty)$, and has the distribution function

$$F(y) = P[Y \leq y] = P[-y \leq Z \leq y] = \frac{1}{\sqrt{2\pi}} \int_{-y}^{y} \exp\left[-\frac{1}{2}y^2\right] dy$$

$$= \frac{2}{\sqrt{2\pi}} \int_{0}^{y} \exp\left[-\frac{1}{2}y^2\right] dy = \sqrt{\frac{2}{\pi}} \int_{0}^{y} \exp\left[-\frac{1}{2}y^2\right] dy. \quad \text{(A.5)}$$

Therefore, the *pdf* of Y

$$f(y) = \frac{dF(y)}{dy} = \sqrt{\frac{2}{\pi}} e^{-\frac{1}{2}y^2} \quad \text{(A.6)}$$

for $y \in [0, \infty)$. By definition, the expectation of Y is

$$E[Y] = \sqrt{\frac{2}{\pi}} \int_{0}^{\infty} ye^{-\frac{1}{2}y^2} dy. \quad \text{(A.7)}$$

But

$$\int_{0}^{\infty} ye^{-\frac{1}{2}y^2} dy = -\lim_{y \uparrow \infty} e^{-\frac{1}{2}y^2} - \left(-e^{-\frac{1}{2}0}\right) = 1, \quad \text{(A.8)}$$

so that

$$E[Y] = \sqrt{\frac{2}{\pi}}. \quad \text{(A.9)}$$

From the definition of Y, it follows that $Y^2 = Z^2$ so that

$$E[Y^2] = E[Z^2] = 1, \quad \text{(A.10)}$$

because $Z \sim N(0, 1)$. As is well known, the variance of Y may be expressed in the form

$$\text{var}[Y] = E[Y^2] - (E[Z])^2, \quad \text{(A.11)}$$

so that

$$\operatorname{var}[Y] = 1 - \frac{2}{\pi} < 1. \tag{A.12}$$

The numerical value of this expression is

$$1 - \frac{2}{\pi} = 0.36338, \tag{A.13}$$

which will be helpful in the numerical evaluation of some formulas to follow.

When formulating a distribution so that it yields nonnegative realizations of random variables in Monte Carlo simulation experiments, one approach would be to consider the random variable defined by

$$W = \mu + \sigma Y = \mu + \sigma |Z|. \tag{A.14}$$

From the above results, it can be seen that

$$E[W] = \mu + \sigma E[Y] = \mu + \sigma \sqrt{\frac{2}{\pi}}. \tag{A.15}$$

Furthermore, the variance of W has the formula

$$\begin{aligned}
\operatorname{var}[W] &= E\left[(W - E[W])^2\right] \\
&= E\left[\left(\mu + \sigma Y - \mu - \sigma\sqrt{\frac{2}{\pi}}\right)^2\right] \\
&= \sigma^2 E\left[\left(Y - \sqrt{\frac{2}{\pi}}\right)^2\right] \\
&= \sigma^2 \operatorname{var}[Y] = \sigma^2\left(1 - \frac{2}{\pi}\right) < \sigma^2.
\end{aligned} \tag{A.16}$$

An advantage of this formulation is that the theoretical expectation and variance are easy to evaluate numerically. Thus, if the random variable defined in (A.14) were used in testing a null hypothesis, as described in previous sections, the formulas for the expectation and variance in (A.15) and (A.16) could be used to estimate the expectation and variance of the random variable W.

From now on attention will be devoted to the folded normal distribution. In the Monte Carlo simulation experiments described in the foregoing sections of this paper a primary task was to compute realizations of a phenotypic random variable W defined by

$$W = |X|, \tag{A.17}$$

where $X = \mu + \sigma Z \sim N(\mu, \sigma^2)$. Some authors call the distribution of the random variable W the folded normal distribution. As a first step in finding the expectation of the random variable W, recall the definition of the function $|x|$ and observe that $|x| = x$ if $x \geq 0$ and $|x| = -x$ if $x \leq 0$. Let A denote the set

$$A = [z \in \mathbb{R} | \mu + \sigma z \geq 0]. \tag{A.18}$$

Equivalently,

$$A = [z \in \mathbb{R} | \mu + \sigma z \geq 0] = \left[z \in \mathbb{R} | z \geq \frac{-\mu}{\sigma} \right]. \tag{A.19}$$

Similarly, let B denote the set

$$B = [z \in \mathbb{R} | \mu + \sigma z \leq 0] = \left[z \in \mathbb{R} | z \leq \frac{-\mu}{\sigma} \right]. \tag{A.20}$$

Then, the expectation of the random variable W may be represented in the form

$$E[W] = E\left[W | Z \geq \frac{-\mu}{2} \right] - E\left[W | Z \leq \frac{-\mu}{2} \right]. \tag{A.21}$$

Observe that

$$\begin{aligned} E\left[W | Z \geq \frac{-\mu}{\sigma} \right] &= z \int_{\frac{-\mu}{\sigma}}^{\infty} (\mu + \sigma z) \frac{1}{\sqrt{2\pi}} e^{-\frac{1}{2}z^2} d \\ &= \mu \left(\int_{\frac{-\mu}{\sigma}}^{\infty} \frac{1}{\sqrt{2\pi}} e^{-\frac{1}{2}z^2} dz \right) + \sigma \int_{\frac{-\mu}{\sigma}}^{\infty} z \frac{1}{\sqrt{2\pi}} e^{-\frac{1}{2}z^2} dz. \end{aligned} \tag{A.22}$$

It can be seen that the coefficient of μ in (A.22)

$$\int_{\frac{-\mu}{\sigma}}^{\infty} \frac{1}{\sqrt{2\pi}} e^{-\frac{1}{2}z^2} dz = 1 - \Phi\left(\frac{-\mu}{\sigma} \right). \tag{A.23}$$

Next, observe that

$$\int_{\frac{-\mu}{\sigma}}^{\infty} z\frac{1}{\sqrt{2\pi}}e^{-\frac{1}{2}z^2}\,dz = -\left(-\frac{1}{\sqrt{2\pi}}e^{-\frac{1}{2}\frac{\mu^2}{\sigma^2}}\right) = \frac{1}{\sqrt{2\pi}}e^{-\frac{1}{2}\frac{\mu^2}{\sigma^2}}. \tag{A.24}$$

Therefore,

$$E\left[W|Z \geq \frac{-\mu}{\sigma}\right] = \mu\left(1-\Phi\left(-\frac{\mu}{\sigma}\right)\right) + \sigma\frac{1}{\sqrt{2\pi}}e^{-\frac{1}{2}\frac{\mu^2}{\sigma^2}}. \tag{A.25}$$

Next, consider

$$E\left[W|Z \leq \frac{-\mu}{\sigma}\right] = \mu\left(\int_{-\infty}^{\frac{-\mu}{\sigma}} \frac{1}{\sqrt{2\pi}}e^{-\frac{1}{2}z^2}\,dz\right) + \sigma\int_{-\infty}^{\frac{-\mu}{\sigma}} z\frac{1}{\sqrt{2\pi}}e^{-\frac{1}{2}z^2}\,dz. \tag{A.26}$$

By definition

$$\int_{-\infty}^{\frac{-\mu}{\sigma}} \frac{1}{\sqrt{2\pi}}e^{-\frac{1}{2}z^2}\,dz = \Phi\left(-\frac{\mu}{\sigma}\right), \tag{A.27}$$

and

$$\int_{-\infty}^{\frac{-\mu}{\sigma}} z\frac{1}{\sqrt{2\pi}}e^{-\frac{1}{2}z^2}\,dz = -\frac{1}{\sqrt{2\pi}}e^{-\frac{\mu^2}{2\sigma^2}}. \tag{A.28}$$

By applying (A.21) and doing the algebra, it follows that

$$\mu_W = E[W] = \mu\left(1-2\sigma^2\Phi\left(-\frac{\mu}{\sigma}\right)\right) + \sigma\sqrt{\frac{2}{\pi}}e^{-\frac{\mu^2}{2\sigma^2}}. \tag{A.29}$$

Because $W^2 = X^2$, it follows that

$$E[W^2] = E[X^2] = \sigma^2 + \mu^2. \tag{A.30}$$

The validity of this equation follows form the equation

$$E[W^2] = \text{var}[W] + (E[W])^2. \tag{A.31}$$

Therefore,

$$\text{var}[W] = \sigma^2 + \mu^2 - (\mu_W)^2. \tag{A.32}$$

From this equation, it can be seen that if $\mu > 0$ and large, then for every fixed $\sigma > 0$

$$\Phi\left(-\frac{\mu}{\sigma}\right) \approx 0 \tag{A.33}$$

and

$$e^{-\frac{\mu^2}{2\sigma^2}} \approx 0. \tag{A.34}$$

Therefore,

$$\mu_W = E[W] \approx \mu, \tag{A.35}$$

and

$$\text{var}[W] \approx \sigma^2 \tag{A.36}$$

is valid. These results are of interest, but under the condition listed above the mean and variance of the folded normal distribution would be near that of the original normal distribution.

In all Monte Carlo simulation experiments reported in the previous sections on testing null hypotheses, σ was chosen as $\sigma = p\mu$, where $0 < p < 1$. In such cases

$$\Phi\left(-\frac{\mu}{\sigma}\right) = \Phi\left(-\frac{1}{p}\right). \tag{A.37}$$

If p is small, then $\Phi\left(-\frac{1}{p}\right) \approx 0$. For example, suppose that $p = 1/4$. Then,

$$\Phi(-4) \tag{A.38}$$

would be small, and if an algorithm were available to evaluate $\Phi(z)$ for any $z \in \mathbb{R}$, then the number in (A.38) could be computed. Next, observe that if

$\sigma = p\mu$, then

$$\exp\left[-\frac{\mu^2}{2p^2\mu^2}\right] = \exp\left[-\frac{1}{2p^2}\right]. \tag{A.39}$$

In particular, if $p = 1/4$, then

$$\exp[-8] = 3.3546 \times 10^{-4}. \tag{A.40}$$

It seems plausible, therefore, that for values of p such that $p \leq 1/2$ the approximations in (A.35) and (A.36) would be near the actual values of μ_W and var$[W]$.

There is another approach to approximating μ_W and var$[W]$ by using Monte Carlo methods and the law of large numbers. For example, let $W_1, W_2, \ldots, W_N$ be independent realizations of the random variable W. Then,

$$\widehat{\mu}_W = \lim_{N\to\infty} \frac{1}{N}\sum_{\nu=1}^{N} W_\nu = \mu_W, \tag{A.41}$$

with probability 1. Similarly,

$$\widehat{E}[W^2] = \lim_{N\to\infty} \frac{1}{N}\sum_{\nu=1}^{N} W_\nu^2 = E[W^2], \tag{A.42}$$

with probability 1. Therefore, if N is large, $N \geq 10{,}000$, then

$$\widehat{\mu}_W \approx \mu_W \tag{A.43}$$

and

$$\widehat{E}[W^2] \approx E[W^2]. \tag{A.44}$$

Hence,

$$\widehat{\text{var}}[W] = \widehat{E}[W^2] - (\widehat{\mu}_W)^2 \approx \text{var}[W]. \tag{A.45}$$

When these approximations are compared with the numerical value of μ_W in (A.29) and that of var$[W]$ in (A.31), an investigator may judge how well the approximations in (A.43) and (A.45) are acceptable.

Lastly, observe that there is another check on the correctness of formulas (A.29) and (A.52). For if $\mu = 0$ and $\sigma = 1$, then from (A.29) and (A.32) it follows that

$$E[W] = \sqrt{\frac{2}{\pi}}, \tag{A.46}$$

and

$$\operatorname{var}[W] = 1 - \frac{2}{\pi}. \tag{A.47}$$

By definition, if $\mu = 0$ and $\sigma = 1$, the random variable W has a half normal distribution with an expectation given by (A.46) and variance (A.47) which match the formulas in (A.9) and (A.12). This demonstration shows that the half normal distribution is a special case of the folded normal distribution, as was expected.

REFERENCES

1. Mode, C. J. and Gallop, R. J. (2008) A review on Monte Carlo simulation methods as they apply to mutation and selection as formulated in Wright–Fisher models of evolutionary genetics. *Math. Biosci.* **211**: 205–225.
2. Mode, C. J., Raj, T. and Sleeman, C. K. (2011) Monte Carlo implementations of two sex density dependent branching processes and their applications in evolutionary genetics. In: C. J. Mode (ed.), Applications of Monte Carlo Methods in Biology, Medicine and Other Fields of Science. INTECH, an Internet company, pp. 273–296.
3. Raj, T., Shulman, J. M., Keenan, B. T., Lori B., Chibnik, L. B., Evans, D. A., Bennett, D. A., Stranger, B. E. and De Jager, P. L. (2012) Alzheimer disease susceptibility loci: evidence for a protein network under natural selection. DOI: 10.1016/j.ajhg.2012.02.022. _2012 The American Society of Human Genetics.
4. Rossin, E. J., Lage, K., Raychaudhuri, S., Xavier, R. J., Tatar, D., *et al.* (2011) Proteins encoded in genomic regions associated with immune-mediated disease physically interact and suggest underlying biology. *PLoS Genet.* 7(1): e1001273. DOI: 10.1371/journal.pgen.1001273 InTech-ISBN 978-953-307-427-6.

Subject Index

Printed in the United States
By Bookmasters